TensorFlow Developer Certificate Guide

Efficiently tackle deep learning and ML problems to ace the
Developer Certificate exam

Oluwole Fagbohun

BIRMINGHAM—MUMBAI

TensorFlow Developer Certificate Guide

Group Product Manager: Ali Abidi

Publishing Product Manager: Anant Jain

Book Project Manager: Farheen Fathima, Aparna Ravikumar Nair

Content Development Editor: Manikandan Kurup

Technical Editor: Sweety Pagaria

Copy Editor: Safis Editing

Proofreader: Safis Editing

Indexer: Manju Arasan

Production Designer: Prashant Ghare, Alishon Mendonca

DevRel Marketing Coordinator: Vinishka Kalra

First published: September 2023

Production reference: 2201023

Published by Packt Publishing Ltd.

Grosvenor House

11 St Paul's Square

Birmingham

B3 1RB, UK.

ISBN 978-1-80324-013-8

www.packtpub.com

The book is dedicated to everyone who has supported me this far especially my parents, my siblings and of course my in-laws. Special thanks goes to my wife and children for their support through those long nights and early morning of writing this book.

Contributors

About the author

Oluwole Fagbohun, a certified TensorFlow developer and machine learning engineer, is currently working as Vice President of Engineering and Environmental Data at ChangeBlock, where he is spearheading the fight against climate change with machine learning. Beyond his role at ChangeBlock, Oluwole is the founder of Readrly an EdTech startup employing AI to enhance children's reading skills, offering engaging and gamified learning experiences.

About the reviewer

Ashish Patel is a seasoned author, data scientist, and researcher with over 11+ years of experience in data science and research. Currently, he holds the position of Sr. AWS AI ML Solution Architect (Chief Data Scientist) at IBM India Pvt Ltd. He has authored *Hands-on Time Series Analysis with Python* and has reviewed 43+ technical books. Ashish's expertise lies in data science, machine learning, deep learning, and MLOps on AWS, with proficiency in AWS SageMaker, AWS Generative AI and ML Services, AWS MLOps, and more. Notably, he has successfully designed and built AI ML project System architecture and built complex enterprise solutions with Red Hat Open Shift and AWS Infrastructure, adhering to the AWS Well-Architected Framework for operational excellence and security. Ashish's passion for AI and ML is evident through his 30+ impactful talks on the subject.

Table of Contents

Part 2 – Image Classification with TensorFlow

5

Image Classification with Neural Networks   95

Part 3 – Natural Language Processing with TensorFlow

11

NLP with TensorFlow   235

Part 4 – Time Series with TensorFlow

12

Introduction to Time Series, Sequences, and Predictions   261

13

Time Series, Sequences, and Prediction with TensorFlow   289

Preface

There is an ever-growing need for deep learning experts across the globe and TensorFlow is a leading framework for building deep learning applications. To meet the growing demand for developers with the requisite skills to build deep learning applications across various domains, the TensorFlow Developer Certificate was established. This 5-hour, hands-on exam is designed to test a developer's foundational knowledge in building, training, saving, and debugging models in computer vision, natural language processing, time series, sequences and predictions.

This book serves as a comprehensive guide in preparing you for this exam. You will begin by understanding the fundamentals of machine learning and the exam in itself. With each chapter, you will acquire new skills with hands-on coding examples and exercises in which you master the science and art of model building for image classification, natural language processing and time series forecasting.

By the end of this book, you will be well equipped with all the nuts and bolts to effectively ace this exam in one sitting.

Who this book is for

This book is for students, data scientists, machine learning engineers and anyone who wants to master how to build deep learning models using TensorFlow and ace the TensorFlow Developer Certificate in the first attempt.

What this book covers

Chapter 1, *Introduction to Machine Learning*, cover, the fundamentals of machine learning, its types, the machine learning life cycle, and applications of machine learning. We will also drill down into what it takes to become a certified TensorFlow developer.

Chapter 2, *Introduction to TensorFlow*, examines the TensorFlow ecosystem after which we set up our work environment. Also, we will take a look at data representation then we will build our hello world model using TensorFlow. We will conclude this chapter by examining how to debug and resolve error messages we come across as we build models.

Chapter 3, *Linear Regression with TensorFlow*, examines how to build linear regression models using TensorFlow. Next, we will explore various evaluation metrics for regression models. We will take matters a step further by building a salary prediction model with TensorFlow and we will close this chapter by mastering how to save and load models.

Chapter 4, Classification with TensorFlow, examines what classification modeling in machine learning is, and discusses different types of classification problems you may encounter in machine learning. Also, we'll examine the various methods of evaluating classification problems and look at how to choose the right classification metrics for your use case. We will close this chapter by looking at a classification problem where we will learn how to build, compile, train, make predictions, and evaluate classification models using TensorFlow.

Chapter 5, Image Classification With Neural Networks, covers the anatomy of neural networks. We'll discuss concepts such as forward propagation, backward propagation, and gradient descent. We will also discuss the moving parts such as the input layer, hidden layers, output layers, activation functions, loss function, and optimizers. We will close this chapter by examining how to build an image classifier using a neural network with TensorFlow.

Chapter 6, Improving the Model, examines various methods of improving the performance of our model. We will discuss the importance of data-centric strategies such as data augmentation and look at various hyperparameters, their impact, and how we can tune them to improve the performance of our models.

Chapter 7, Image Classification with Convolutional Neural Networks, introduces **convolutional neural networks (CNNs)**. We'll see how CNNs change the game in image classification tasks by exploring its anatomy, exploring concepts like convolutions and pooling operations. We will examine the challenges developers face when working with real world images. We will close the chapter by seeing CNNs in action as we apply them to classifying weather image data.

Chapter 8, Handling Overfitting, examines overfitting in greater details. We will examine what it is and why it occurs in real-world use cases. We will then go a step further by exploring various ways to overcome overfitting such as dropout regularization, early stopping and L1 & L2 regularization. We will put these ideas to the test using the weather image dataset case study to help us cement our understanding of these concepts.

Chapter 9, Transfer Learning, introduces the concept of transfer learning and we will discuss where and how we can apply transfer learning. We will also examine some best practices around applying transfer learning in our workflow. To close this chapter, we will build a real-world image classifier using pretrained models from TensorFlow.

Chapter 10, Introduction to Natural Language Processing, introduces natural language processing. We will discuss challenges around working with text data, and how we can move from language to vector representations. We will cover foundational ideas around text preprocessing and data preparation techniques such as tokenization, padding, sequencing, and word embedding. We will step things up by examining how to visualize word embeddings using TensorFlow's projector. To close this chapter, we will build a sentiment classifier using TensorFlow.

Chapter 11, NLP with TensorFlow, goes deeper into the challenges of modeling text data. We will introduce **recurrent neural networks (RNNs)** and their variants, **long short-term memory (LSTM)** and **gated recurrent units (GRU)**. We'll see how they are uniquely tailored for handling sequential data such as text and time series data. We will apply these models in building a text classifier. We will also see how to apply pretrained word embeddings in this chapter. To close this chapter, we will see how we can build a children's story generator using LSTMs.

Chapter 12, Introduction to Time Series, Sequences and Predictions, introduces time series data and examines the unique nature of time series data, its core characteristics, types and application of time series data. We will discuss some of the challenges with modeling time series data and examine a number of solutions. We will see how to prepare time series data for forecasting using utilities in TensorFlow, and then we will apply both statistical methods and machine learning techniques to forecast sales data for a fictional super store.

Chapter 13, Time Series, Sequences and Prediction with TensorFlow, discusses how to use in-built and custom learning rate schedulers in TensorFlow, we will also see how to apply lambda layers. We will see how to build time series forecasting models using RNNs, LSTMs, CNNs and CNN-LSTM networks. We will round off this chapter and the book by working through a problem where we will collect realworld closing stock prices and build a forecasting model.

To get the most out of this book

To get the most out of this book you need to be proficient in Python and need to have a desire to learn all things in machine learning. We'll be using Google Colab in this book but you can use your preferred IDE with the following requirements:

Software/hardware covered in the book	Operating system requirements
Python 3.9.2	
`tensorflow==2.13.0`	
`numpy==1.24.3`	
`pandas==2.0.3`	Windows, macOS, or Linux
`Pillow==10.0.0`	
`scipy==1.10.1`	
`tensorflow-datasets==4.9.2`	

If you are using the digital version of this book, we advise you to type the code yourself or access the code from the book's GitHub repository (a link is available in the next section). Doing so will help you avoid any potential errors related to the copying and pasting of code.

Download the example code files

You can download the example code files for this book from GitHub at `https://github.com/PacktPublishing/TensorFlow-Developer-Certificate-Guide`. If there's an update to the code, it will be updated in the GitHub repository.

We also have other code bundles from our rich catalog of books and videos available at `https://github.com/PacktPublishing/`. Check them out!

Conventions used

There are a number of text conventions used throughout this book.

`Code in text`: Indicates code words in text, database table names, folder names, filenames, file extensions, pathnames, dummy URLs, user input, and Twitter handles. Here is an example: "This metadata information is stored in the `info` variable "

A block of code is set as follows:

```
import numpy as np
from numpy import
```

When we wish to draw your attention to a particular part of a code block, the relevant lines or items are set in bold:

```
(<tf.Tensor: shape=(2, 3), dtype=int64,
    numpy=array([[10, 11, 12], [ 3, 4, 5]])>,
 <tf.Tensor: shape=(2,),
    dtype=int64, numpy=array([13, 6])>)
```

Bold: Indicates a new term, an important word, or words that you see onscreen. For instance, words in menus or dialog boxes appear in **bold**. Here is an example: "**Convolutional neural networks (CNNs)** are the go-to algorithms when it comes to image classification."

Tips or important notes
Appear like this.

Get in touch

Feedback from our readers is always welcome.

General feedback: If you have questions about any aspect of this book, email us at `customercare@` `packtpub.com` and mention the book title in the subject of your message.

Errata: Although we have taken every care to ensure the accuracy of our content, mistakes do happen. If you have found a mistake in this book, we would be grateful if you would report this to us. Please visit `www.packtpub.com/support/errata` and fill in the form.

Piracy: If you come across any illegal copies of our works in any form on the internet, we would be grateful if you would provide us with the location address or website name. Please contact us at `copyright@packt.com` with a link to the material.

If you are interested in becoming an author: If there is a topic that you have expertise in and you are interested in either writing or contributing to a book, please visit `authors.packtpub.com`.

Share Your Thoughts

Once you've read *TensorFlow Developer Certificate Guide*, we'd love to hear your thoughts! Scan the QR code below to go straight to the Amazon review page for this book and share your feedback.

`https://packt.link/r/1-803-24013-X`

Your review is important to us and the tech community and will help us make sure we're delivering excellent quality content.

Download a free PDF copy of this book

Thanks for purchasing this book!

Do you like to read on the go but are unable to carry your print books everywhere?

Is your eBook purchase not compatible with the device of your choice?

Don't worry, now with every Packt book you get a DRM-free PDF version of that book at no cost.

Read anywhere, any place, on any device. Search, copy, and paste code from your favorite technical books directly into your application.

The perks don't stop there, you can get exclusive access to discounts, newsletters, and great free content in your inbox daily

Follow these simple steps to get the benefits:

1. Scan the QR code or visit the link below

https://packt.link/free-ebook/978-1-80324-013-8

2. Submit your proof of purchase
3. That's it! We'll send your free PDF and other benefits to your email directly

Part 1 –
Introduction
to TensorFlow

n this part of the book, you will learn the foundations of machine learning and deep learning required to succeed in this exam. You will learn to read data from different sources and in different formats. You will also learn how to build a regression and classification model, debug, save, load the model, and also make predictions with real world data.

This section comprises of the following chapters:

- *Chapter 1, Introduction to Machine Learning*
- *Chapter 2, Introduction to TensorFlow*
- *Chapter 3, Linear Regression with TensorFlow*
- *Chapter 4, Classification with TensorFlow*

1

Introduction to Machine Learning

There has never been a more exciting time to be a deep learning expert. With the advent of super-fast computers, open source algorithms, well-curated datasets, and affordable cloud services, deep learning experts are armed with the requisite skills to build amazing and impactful applications across all domains. Computer vision, natural language processing, and time series analysis are just a few of the areas where deep learning experts can make a real impact. Anyone with the right skills can build a groundbreaking application and perhaps become the next Elon Musk. For this to happen, adequate knowledge of deep learning frameworks such as TensorFlow is required.

The TensorFlow Developer Certificate aims to build a new generation of deep learning experts who are already in high demand across all fields. Hence, joining this club equips you with the required expertise to start your journey as a deep learning expert and also presents you with a certificate to show for your weeks, months, or years of hard work.

We will begin this chapter with a high-level introduction to **machine learning** (**ML**), after which we will examine the different types of ML approaches. Next, we will drill down into the ML life cycle and use cases (we will cover a few hands-on implementations in subsequent chapters). We conclude this chapter by introducing the TensorFlow Developer Certificate and examining the anatomy of the core components needed to ace the exam. By the end of this chapter, you should be able to clearly explain what ML is and have gained a foundational understanding of the ML life cycle. Also, after this chapter, you will be able to differentiate between different types of ML approaches and clearly understand what the TensorFlow Developer Certificate exam is all about.

In this chapter, we will cover the following topics:

- What is ML?
- Types of ML algorithms
- The ML life cycle

- Exploring ML use cases
- Introducing the learning journey

What is ML?

ML is a subfield of **artificial intelligence (AI)** in which computer systems learn patterns from data to perform specific tasks or make predictions on unseen data without being explicitly programmed. In 1959, Arthur Samuel defined ML as a *"field of study that gives computers the ability to learn without being explicitly programmed to do so."* To give clarity to the definition given to us by Arthur Samuel, let us unpack it using a well-known use case of ML in the banking industry.

Imagine we work in the ML team of a Fortune 500 bank in the heart of London. We are saddled with the responsibility of automating the fraud detection process as the current manual process is too slow and costs the bank millions of pounds sterling, due to delays in the transaction processing time. Based on the preceding definition, we request historical data of previous transactions, containing both fraudulent and non-fraudulent transactions, after which we go through the ML life cycle (which we will cover shortly) and deploy our solution to prevent fraudulent activities.

In this example, we used historical data that provides us with the features (independent variables) needed to determine the outcome of the model, which is generally referred to as the target (dependent variable). In this scenario, the target is a fraudulent or non-fraudulent transaction, as shown in *Figure 1.1.*

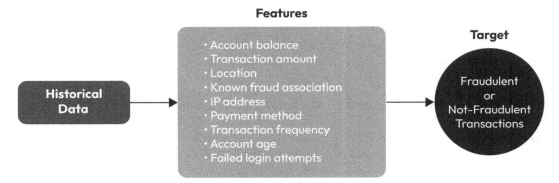

Figure 1.1 – A flowchart showing the features and the target in our data

In the preceding scenario, we were able to train a model with historical data made up of features and the target to generate rules that are used to make predictions on unseen data. This is an example of what ML is all about – the ability to empower computers to make decisions without explicit programming. In classic programming, as shown in *Figure 1.2*, we feed in the data and some hardcoded rules – for example, the volume of a daily transaction to determine whether it is fraudulent or not. If a customer goes above this daily limit, the customer's account gets flagged, and a human moderator will intervene to decide whether the transaction was fraudulent or not.

Figure 1.2 – A traditional programming approach

This approach will soon leave the bank overwhelmed with unhappy customers constantly complaining about delayed transactions, while fraudsters and money launderers will evade the system by simply limiting their transactions within the daily permissible limits defined by the bank. With every new attribute, we would need to update the rules. This approach quickly becomes impractical, as there will always be something new to update to make the system tick. Like a house of cards, the system will eventually fall apart, as such a complex problem with continuously varying attributes across millions of daily transactions may be near impossible to explicitly program.

Thankfully, we do not need to hardcode anything. We can use ML to build a model that can learn to identify patterns of fraudulent transactions from historical data, based on a set of input features in it. We train our model using labeled historical data of past transactions that contain both fraudulent and non-fraudulent transactions. This allows our model to develop rules based on the data, as shown in *Figure 1.3*, which can be applied in the future to detect fraudulent transactions.

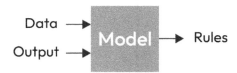

Figure 1.3 – An ML approach

The rules generated by examining the data are used by the model to make new predictions to curb fraudulent transactions. This paradigm shift differs from traditional programming where applications are built using well-defined rules. In ML-powered applications, such as our fraud detection system, the model learns to recognize patterns and create rules from the training data; it then uses these rules to make predictions on new data to efficiently flag fraudulent transactions, as shown in *Figure 1.4*:

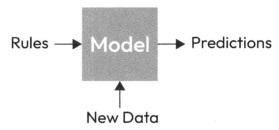

Figure 1.4 – An ML model uses rules to make predictions on unseen data

In the example we just examined, we can see from *Figure 1.1* that our training data is usually structured in a tabular form, made up of numerical values such as transaction amount and frequency of transactions, as well as categorical variables such as location and payment type. In this type of data representation, we can easily identify both the features and the target. However, what about textual data from social media, images from our smartphones, video from streaming movies, and so on, as illustrated in *Figure 1.5*? How do we approach these types of problems where the data is unstructured? Thankfully, we have a solution, which is called **deep learning**.

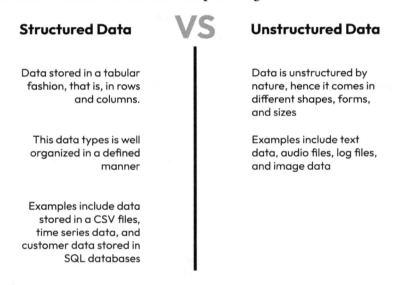

Figure 1.5 – An illustration of structured and unstructured data types

Deep learning is a subset of ML that mimics the human brain by using complex hierarchical models that are composed of multiple processing layers. The buzz around deep learning lies around the state-of-the-art performance recorded by deep learning algorithms over the years across many real-world applications, such as object detection, image classification, and speech recognition, as deep learning algorithms can model complex relationships in data. In *Sections 2* and *3*, we will discuss deep learning in greater detail and see it in action for image and text applications, respectively. For now, let us probe further into the world of ML by looking at the types of ML algorithms.

Types of ML algorithms

In the last section, we looked at what ML is, and we examined a use case where we had labeled data. In this section, we will look at the four main types of ML approaches to give us a base understanding of what each type does, and where and how they can be applied. The four types of ML algorithms are as follows:

- Supervised learning
- Unsupervised learning

- Semi-supervised learning
- Reinforcement learning

Let's examine the four types of ML methods, which will serve as a building block for later chapters.

Supervised learning

In **supervised learning**, an ML model is trained using data made of features and a target. This enables the model to learn the underlying relationship in the data. After training, the model can use its newfound knowledge to make predictions on unseen data. For example, say you want to buy a house. You would consider factors such as the location of the house, the number of rooms, the number of bathrooms, whether it has a garden, and the type of property; these factors you would consider as the features, while the price of the house is the target. Perhaps after the TensorFlow exam, you can roll up your sleeves, scrape some housing data, and train a model to predict house prices based on these features. You can use your house prediction model to compare prices on real estate websites to close a good deal for yourself.

There are two types of supervised learning – regression and classification. In a regression task, the label is a numeric value, just like the example we gave previously in which the target is the predicted price of the house. Conversely, in a classification task, the label is a category, just like the fraud example we discussed previously, in which the goal was to detect whether a transaction was fraudulent or not fraudulent. In a classification task, the model will learn to categorize the target as classes.

When dealing with a classification task made up of two classes (for example, fraudulent and non-fraudulent transactions), it is referred to as binary classification. When there are more than two categories (for example, the classification of different car brands), it is referred to as multi-class classification.

Figure 1.6: An example of multi-label classification, where the
model identifies multiple subjects within an image

Multi-label classification is another type of classification, and it is used for image tagging in social media applications such as Facebook and Instagram. Unlike binary and multi-class classification tasks in which we have one target per instance, in multi-label classification our model identifies multiple targets for each instance, as shown in *Figure 1.6*, where our model identifies a girl, a dog, a boy, and a cat in the given photo.

Unsupervised learning

Unsupervised learning is the opposite of supervised learning. In this case, there is no labeled data. The model will have to figure things out by itself. Here, an unsupervised learning algorithm is provided with data, and we expect it to extract meaningful insights from the unlabeled data, by identifying patterns within the data without relying on predefined targets – for example, an image you work on for a large retail store that aims to segment its customers for marketing purposes. You are given data containing the demographics and spending habits of the store's customers. By employing an unsupervised ML model, you successfully cluster customers with similar characteristics into distinct customer segments. The marketing team can now create a tailored campaign targeted at each of the customer segments identified by the model, potentially leading to a higher conversion rate from the campaign.

Semi-supervised learning

Semi-supervised learning is a combination of supervised and unsupervised learning. In this case, there is some data with labels (i.e., it has both features and a target) and the rest is without labels. In such scenarios, the unlabeled data is usually the dominant group. In this case, we can apply a combination of unsupervised and supervised learning methods to generate optimal results, especially when the cost implication and time required to manually label the data may not be practical. Here, the model takes advantage of the available label data to learn the underlying relationships, which it then applies to the unlabeled data.

Imagine you work for a big corporation that collects a large volume of documents that we need to classify and send to the appropriate department (i.e, finance, marketing, and sales) for effective document management, and only a small number of the documents are labeled. Here, we apply semi-supervised learning, train the model on the labeled document, and apply the learned patterns to classify the remaining unlabeled documents.

Reinforcement Learning

In **reinforcement learning**, unlike supervised learning, the model does not learn from training data. Instead, it learns from its interactions within an environment; it gets rewarded for making the right decisions and punished for making wrong choices. It is a trial-and-error learning approach in which the model learns from its past experience to make better decisions in the future. The model aims to maximize reward. Reinforcement learning is applied in self-driving cars, robotics, trading and finance, question answering, and text summarization, among other exciting use cases.

Now that we can clearly differentiate between the different types of ML approaches, we can look into what the core components of an ML life cycle are and what steps we should take, from the birth of a project to its application by end users. Let us take a look at the ML life cycle.

ML life cycle

Before embarking on any ML project, we must take into account some key components that can determine whether our project will be successful or not. And this is important because as data professionals who want to build and implement successful ML projects, we need to understand how the ML life cycle works. The ML life cycle is a sensible framework to implement an ML project, as shown in *Figure 1.7*:

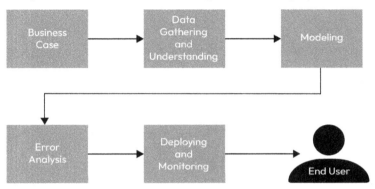

Figure 1.7 – The ML life cycle

Let's look at each of these in detail.

The business case

Before unleashing state-of-the-art models on any problem, it is imperative you take time to sit with stakeholders to clearly understand the business objectives or the pain points to be resolved, as without clarity, the entire process will almost definitely fail. It is always important to keep in mind that the goal of the entire process is not to test a new breakthrough model you have been itching to try out but to solve a pain point, or create value for your company.

Once we understand the problem, we can categorize the problem as either a supervised or unsupervised learning task. This phase of an ML life cycle is all about asking the right questions. We need to sit with the concerned team to determine what the key metrics that would define the project as a success are. What resources are required in terms of budget, manpower, compute, and the project timeline? Do we have the domain understanding or do we need an expert's input into defining and understanding the underlying factors and goals that will define the project's success? These are some of the questions we should ask as data professionals before we embark on a project.

For the exam, we will need to understand the requirements of each question before we tackle them. We will discuss a lot more about the exam before we conclude this chapter.

Data gathering and understanding

When all the requirements are detailed, the next step is to collect the data required for the project. In this phase, we would first determine what type of data we will collect and where we will collect it from. Before we embark on anything, we need to ask ourselves whether the data is relevant – for example, if we collect historical car data from 1980, would we be able to predict the price of a car in 2022? Would data be made available by stakeholders, or would we be collecting it from a database, **Internet of Things (IoT)** devices, or via web scraping? Would there be any need for the collection of secondary data for the task at hand? Also, we would need to establish whether the data will be collected all at once or whether it will be a continuous process of data collection. Once we have collected the data needed for the project, we would then examine the data to get an understanding of it.

Next, we would examine the data to see whether the data collected is in the right format. For example, if you collect car sales data from multiple sources, one source may calculate a car's mileage in kilometers per hour and another source could use miles per hour. Also, there could be missing values in some of the features, and we might also encounter duplicates, outliers, and irrelevant features in the data we collected. During this phase, we would carry out data exploration to gain insights into the data, and data preprocessing to handle various issues such as formatting problems, missing values, duplicates, removal of irrelevant features, and handling outliers, imbalanced data, and categorical features.

Modeling

Now that we have a good understanding of the business needs, we have decided on the type of ML problem that we will address, and we also have good-quality data after completing our preprocessing step. We will split our data into a training split and keep a small subset of the test to evaluate the model's performance. We will train our model to understand the relationship between the features and the target variable using our training set. For example, we could train our fraud detection model on historical data provided by the bank and test it out with our hold out (test set) to evaluate our model's performance before deploying it for use. We go through an iterative process of fine-tuning our model hyperparameters until we arrive at our optimal model.

Defining whether the modeling process is a success or not is tied to the business objective, since achieving a high accuracy of 90 percent would still leave room for a 10 percent error, which could be decisive in high-stake domains such as healthcare. Imagine you deploy a model for early-stage cancer detection with an accuracy of 90 percent, which means the model would likely fail once for every 10 people; in 100 tries, it could fail about 10 times, and it could misclassify someone with cancer as healthy. This could lead to the individual not only failing to seek medical advice but also to an untimely demise. Your company could get sued and the blame would fall in your lap. To avoid situations like this, we need to understand what metrics are important for our project and what we should be less strict with. It is also important to address factors such as class imbalance, model interpretability, and ethical implications.

There are various metrics that are used to evaluate a model, and the type of evaluation depends on the type of problem we will handle. We will discuss regression metrics in *Chapter 3, Linear Regression with TensorFlow,* and classification metrics in *Chapter 4, Classification with TensorFlow,.*

Error analysis

We are not ready for deployment yet. Remember the 10 percent data that could tank our project? We will address that here. We perform an error analysis to identify the misclassified labels to identify why the model missed them. Do we have enough representative samples of these misclassified labels in our training data? We would have to determine whether we need to collect more data to capture these cases where the model failed. Can we generate synthetic data to capture the misclassified labels? Or was the misclassified data down to the wrong labeling?

Wrongly labeled data can hamper the performance of a model, as it will learn incorrect relationships between the features and target, resulting in poor performance on unseen data, making the model unreliable and the entire process a waste of resources and time. Once we resolve these questions and ensure accurate labels, we need to retrain and reevaluate our model. These steps are continuous until the business objective is achieved, and then we can proceed to deploy our model.

Model deployment and monitoring

After resolving the issues identified in the error analysis step, we can now deploy our model to production. There are various methods of deployment available. We could deploy our model as a web service, on the cloud, or on edge devices. Model deployment can be challenging as well as exciting because the entire point of building and training a model is to allow end users to apply it to solve a pain point. Once we deploy our model, we also monitor the model to ensure that the overall objectives of the business are continually achieved, and even the best-performing models can begin to underperform over time due to concept drift and data drift. Hence, after deploying our model, we cannot retire to some island. We need to continuously monitor our model and retrain the model when needed in order to ensure it continues to perform optimally.

We have now gone through the full length of the ML life cycle at a high level. Of course, there is a lot more that we can talk about in greater depth, but this is out of the scope of this exam. Hence, we will now switch our focus to looking at a number of exciting use cases where ML can be applied.

Exploring ML use cases

Beyond the car price prediction and fraud detection use cases, let's look at some exciting applications of ML here. Perhaps it will get you fired up for both the exams and inspire you to build something spectacular in your ML journey.

Healthcare

HearAngel uses AI to automatically prevent hearing loss for headphone users by tracking users' exposure to unhealthy levels of sounds from the headphones. Insitro uses knowledge of ML and biology for drug discovery and development. Other use cases of ML in healthcare include smart record keeping, data collection, disease outbreak forecasting, personalized medicine, and disease identification.

The retail industry

ML engineers are revolutionizing the retail industry by deploying models to enhance customer experience and improve profitability. This is achieved by optimizing assortment planning, predicting customer behavior, the provision of virtual assistants, inventory optimization, tracking customers' sentiments, price optimization, product recommendation, and customer segmentation, among other use cases. In the retail industry, ML engineers provide value by automating cumbersome manual processes.

The entertainment industry

The entertainment industry currently applies ML/AI in automatic script and lyric generation. Yes, currently there are movie script-writing ML models. One such movie is a short science fiction movie called *Sunspring* (`https://arstechnica.com/gaming/2021/05/an-ai-wrote-this-movie-and-its-strangely-moving/`). Also, ML/AI is used for automatic caption generation, augmented reality, game development, target marketing, sentiment analysis, movie recommendation, and sales forecasting, among others. So, if you plan to carve a niche in this industry as an ML engineer, there is a lot you can do, and surely you can also come up with some new ideas.

Education

Readrly utilizes deep learning techniques to create personalized children's stories, enhancing the learning experience for young readers. By tailoring stories to each child's interests and skill level, Readrly supports children's reading development in a fun and engaging way.

Agriculture

There are numerous use cases of ML in agriculture. ML/AI can be used for price forecasting, disease detection, weather prediction, yield mapping, soil and crop health monitoring, and precision farming, among others.

Here, we covered some use cases of ML. However, the exciting part is that, as an ML/DL engineer, you will be able to apply your knowledge to any industry as long as data is available. And that is the awesomeness of being an ML engineer – there are no restrictions. We have covered a lot already in this chapter; however, we have one more section to go, and it is a very important one, as it centers on the exam itself. Let's jump in.

Introducing the learning journey

The TensorFlow Developer Certificate exam was designed and developed by Google to assess data professionals' expertise in model building and training deep learning models with TensorFlow. The exams enable data professionals to showcase their skills in solving real-world problems with ML/DL, as shown in *Figure 1.8*.

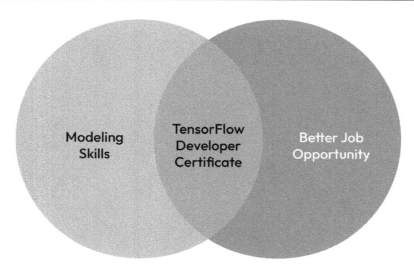

Figure 1.8 – The goal of the exam

Let's dig deeper into why you should take this exam.

Why take the exam?

One of the most compelling reasons why you should take the TensorFlow Developer Certificate is that it can help you get a job. The global AI market is expected to grow exponentially, reaching $2 trillion by 2030 according to a report by Statista (`https://www.statista.com/statistics/1365145/ artificial-intelligence-market-size/#:~:text=The%20market%20for%20 artificial%20intelligence,nearly%20two%20trillion%20U.S.%20dollars`).

This rapid growth is being driven by continued advancements in areas such as autonomous vehicles, image recognition, and natural language processing, powering a new wave of applications across a wide range of industries. This growth is expected to create an increased demand for deep learning specialists who can build cutting-edge ML solutions.

In light of this development, recruiters and hiring managers are on the lookout for skilled candidates who can build deep learning models with TensorFlow, and this certificate can help you stand out from the crowd. To further accelerate your job-hunting, Google has put together *TensorFlow Certificate Network*, which is an online database of certified TensorFlow developers across the globe, as shown in *Figure 1.9*. Hiring managers can easily find suitable candidates to build their machine learning and deep learning solutions using a range of filters, such as location and years of experience, as well as verifying a candidate based on their name.

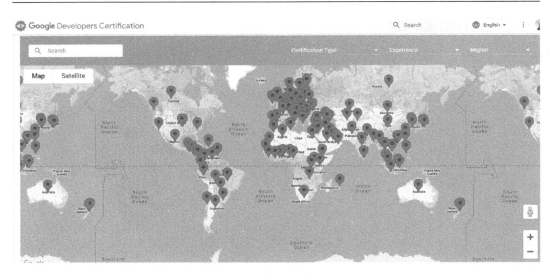

Figure 1.9 – A map display of TensorFlow-certified developers

In addition to helping you get a first job, the TensorFlow Developer Certificate can also help you advance your career. If you're already working with TensorFlow, the certificate can help you demonstrate your expertise to your employer. This can lead to promotions and raises.

Now that we have looked at some of the reasons why you should take the exam, the next logical step is to look at what the exam is all about. Let us do that.

What is the exam all about?

If you're planning on becoming a certified TensorFlow developer, there are a few things you'll need to know. Here's a quick rundown of what you need to know to ace the TensorFlow Developer Certificate exam:

- TensorFlow developer skills
- Building and training neural network models using TensorFlow 2.x
- Image classification
- **Natural language processing (NLP)**
- Time series, sequences, and predictions

You can find the complete exam details here: `https://www.tensorflow.org/static/extras/cert/TF_Certificate_Candidate_Handbook.pdf`. However, this book covers each section of the exam in detail to help ensure success. The exam costs $100, but there is an option to apply for a stipend that, if approved, will allow you to only pay half the exam fee. The stipend must be used within 90 days of receiving it and is only valid for one attempt. To apply for the stipend, you must provide information about yourself, why you need the stipend, and your portfolio projects with TensorFlow if you have any. You can find more information about how to access the TensorFlow Education Stipend using this link: `https://www.tensorflow.org/static/extras/cert/TF_Education_Stipend.pdf`.

We have now covered what and why. Now, let us look at how you can ace the exam.

How to ace the exam

If you're looking to become a certified TensorFlow developer, there are a few things you should know. First, you should be proficient in Python. Second, you'll need to have a strong understanding of ML concepts and be able to use TensorFlow to build and train deep learning models with TensorFlow. If you're not already familiar with Python programming, then *Modern Python Cookbook – Second Edition* by Steven F Lott is a good place to start.

Here are some tips to help you ace the TensorFlow Developer Certificate exam:

- **Review the course material**: Before taking the exam, be sure to review the materials for every topic in the TensorFlow candidate handbook in detail. Pay special attention to building and training models, since the exam is hands-on.

- **Model building**: In addition to reviewing the course material, it's also important to get some hands-on experience with TensorFlow. Experiment with building models, covering each section of the exam requirement. This book will get you started with the core fundamentals of ML and walk you through each section of the exam in a hands-on manner, ensuring that you can comfortably build and train all sorts of models with TensorFlow – from simple linear models to complex neural networks.

- **Understand the exam format**: The exam is unlike many other exams. It is a five-hour coding exam with questions covering each section of the exam that we outlined. You will be given a task and asked to write code to solve it in PyCharm. So, you will need to spend some time mastering how to build, train, and save models in PyCharm before the exams. The exam is an open-book one, so you are allowed to use any resource you want during the exam.

- **Practice, practice, practice**: One of the best ways to prepare for the exam is to practice solving questions. You will find lots of hands-on practice questions in every chapter of this book, along with code files in the book's GitHub repository. Additionally, you will find lots of datasets on the TensorFlow website and Kaggle.

Once you've completed this book, you should be ready to take the TensorFlow Developer Certificate exam.

When to take the exam

Depending on your experience, you may need more or less time to prepare for the exam. If you are familiar with TensorFlow already, with hands-on model-building skills, you may take between three weeks to two months to prepare for the exam. However, if you are completely new to TensorFlow, it is advisable to look at around six months to thoroughly prepare for this exam, as stipulated on the exam website. However, these rules are not cast in stone. Everyone is different, so go at your own pace.

Exam tips

After you sign up for the exam, you can take it within a six-month period. It is okay to set a target day for your exams early enough. The exam takes place in PyCharm, so you will need to take a few days or weeks to get used to PyCharm (if you're not familiar with it) before the exam day. Here is an excellent video tutorial by Jeff Henton to help you set up your PyCharm environment: `https://www.youtube.com/watch?v=zRY5lx-So-c`. Ensure you install the stipulated version of PyCharm. You can also learn more about setting up the exam environment using this link: `https://www.tensorflow.org/static/extras/cert/Setting_Up_TF_Developer_Certificate_Exam.pdf`.

Before the exam, it helps to have a clearly planned study routine in which you can cover the outlined syllabus for the exam. This book will help you on this journey, so you should pay attention to the next chapters, as we will start coding and tackling the core components, which make up the exam henceforth. On exam day, I would advise you to find a quiet, comfortable space to take this exam. Ensure you are well rested and not going in exhausted, as the exam is five hours long. Also, try out your PyCharm and internet connectivity. Don't panic, and read the questions thoroughly to have a clear understanding of what is required of you. Start from questions 1 to 5. Since the questions get more difficult, it is better to get the easy ones out of the way quickly and tackle the more difficult ones afterward.

However, you should pace yourself correctly. Your saved model will be graded each time you submit it, and you are allowed to submit as many times as you want within the stipulated 5-hour time frame until you achieve an optimal result. You can also run your model in Colab if this enables you to work faster, especially if you have a model running in PyCharm. Colab provides you with free GPU access to train your model. The exam will be graded only in PyCharm, so bear this in mind. Ensure you save the model you train in Colab, and move it to the directory where you are stipulated to save the model for the exam.

If you need help, you can use Stack Overflow. You can also look through the code we use in this book, or any other material you use to prepare for your exam. However, if a question proves too difficult, move on to other questions. When you are done, you can return to the difficult one and work your way through it to avoid losing all your time on a difficult question. Also, you are allowed to submit multiple times, so work on improving your models until you attain optimal performance.

What to expect after the exam

The exam ends after five hours exactly, but you can submit it before that. After which, you will receive a congratulatory email if you have passed. After passing this exam, you will now be a member of the Google TensorFlow Developer community, opening more doors of opportunity for yourself. Assuming you pass (and I hope you do), you will get your certificate in about a week, which will look like *Figure 1.10*, and you will be added to the Google TensorFlow community in about two weeks. The certificate is valid for a period of three years.

Figure 1.10 – TensorFlow Developer Certificate

Now, you know the topics, the time frame, the cost, how to prepare, what to do on exam day, and what to expect after the exam. With this, we have come to the end of this chapter. We have covered a lot of theory in this chapter, which will serve as the basis for the work we will do together in the upcoming chapters.

Summary

This chapter provided an overview of ML, deep learning, and the types of ML approaches. It also covered the ML life cycle and various ML use cases across different domains. We looked at a high-level overview of the TensorFlow Developer Certificate, along with information on the components of the exam and how to prepare for it. At the end of this chapter, you should have a good foundational understanding of what ML is and its types. You should now be able to determine which problems are ML-based problems and those that require classic programming. You should also be able to unpack ML problems into different types and be familiar with what it takes to prepare for the TensorFlow Developer Certificate exam by Google.

In the next chapter, we will look at what TensorFlow is, set up our environment, and start coding our way to the end of this book.

Questions

Let's test what we learned in this chapter:

1. What is ML?
2. What is deep learning?
3. What are the types of ML?
4. What are the steps in the ML life cycle?
5. What is the TensorFlow Developer Certificate about?
6. What are the core areas of the exam?

Further reading

To learn more, you can check out the following resources:

-   *TensorFlow Developer Certificate overview*: https://www.tensorflow.org/certificate
-   *Hands-On Machine Learning with scikit-learn and Scientific Python: Toolkits* by Amr, T., Packt Publishing
-   *Deep Learning: Methods and Applications* by Li Deng, and Dong Yu: https://doi.org/10.1561/2000000039
-   *Movie written by algorithm turns out to be hilarious and intense*: https://arstechnica.com/gaming/2021/05/an-ai-wrote-this-movie-and-its-strangely-moving/.
-   *Python Machine Learning – Third Edition* by Sebastian Raschka and Vahid Mirjalili

2
Introduction to TensorFlow

Before the era of TensorFlow, the landscape of deep learning was markedly different. Data professionals had fewer comprehensive tools to aid in the development, training, and deployment of neural networks. This posed challenges in experimenting with various architectures and tuning model settings to solve complex tasks, as data experts often had to construct their models from scratch. The process was time-consuming, with some experts spending days or even weeks developing effective models. Another bottleneck was the difficulty in deploying trained models, which made the practical application of neural networks challenging during those early days.

But today, everything has changed; with TensorFlow, you can do lots of amazing things. In this chapter, we will begin by examining the TensorFlow ecosystem and discussing, at a high level, the various components relevant to building state-of-the-art applications with TensorFlow. We will begin our journey by setting up our workspace to meet the requirements of the exam and our upcoming experiments. We will also learn what TensorFlow is all about, understand the concept of tensors, explore basic data representation and operations in TensorFlow, and build our first model using this powerful tool. We will conclude this chapter by looking at how to debug and solve error messages in TensorFlow.

By the end of this chapter, you will understand the basics of TensorFlow, including what tensors are and how to perform basic data operations with them. You will be equipped to confidently build your first model with TensorFlow and debug and solve any error messages that might arise in the process.

In this chapter, we will cover the following topics:

- What is TensorFlow?
- Setting up our environment
- Data representation
- Hello World in TensorFlow
- Debugging and solving error messages

Technical requirements

We will be using **Google Colaboratory (Google Colab)** notebooks as our work environment as it is a free, cloud-based Jupyter Notebook service that is easy to use and provides us with **graphics processing unit (GPU)** and **tensor processing unit (TPU)** backends. We will be using Google Colab to run the coding exercise, which requires `python >= 3.8.0` along with the following packages, which can be installed using the `pip install` command:

- `tensorflow>=2.7.0`
- `tensorflow-datasets==4.4.0`
- `pillow==8.4.0`
- `pandas==1.3.4`
- `numpy==1.21.4`
- `scipy==1.7.3`

The code bundle for this book is available at the following GitHub link: `https://github.com/PacktPublishing/TensorFlow-Developer-Certificate`. Also, solutions to all exercises can be found in the GitHub repo itself. If you are new to Google Colab, here is a great resource to get you started quickly: `https://www.youtube.com/watch?v=inN8seMm7UI&list=PLQY2H8rRoyvyK5aEDAI3wUUqC_F0oEroL`.

What is TensorFlow?

In the last chapter, we examined the different types of applications we could build with our knowledge of **machine learning (ML)**, from chatbots to facial recognition systems and from house price prediction to detecting fraud in the banking industry – these are some of the exciting applications we can build using deep learning frameworks such as TensorFlow. The question we would logically ask is what exactly is TensorFlow? And why should we learn it at all?

TensorFlow is an open source end-to-end framework for building deep learning applications. It was developed by a team of data professionals at Google in 2011 and made openly available in 2015. TensorFlow is a flexible, scalable solution that enables us to build models with ease using the Keras API. It allows us to access a large array of pretrained deep learning models, thus making it the framework of choice for many data professionals in the industry and academia. Currently, TensorFlow is used at powerhouses such as Google, DeepMind, Airbnb, Intel, and so many more companies.

Today, with TensorFlow, you can easily train a deep learning model on a single PC, using a cloud service such as AWS, or using distributed training with a cluster of computers. Model building is just a part of what data professionals do; what about visualizing, deploying, and monitoring our models? TensorFlow has a wide range of tools to cater to these processes, such as TensorBoard, TensorFlow lite, TensorFlow.js, TensorFlow Hub, and **TensorFlow Extended (TFX)**. These tools enable data professionals to build and deploy scalable, low-latency, ML-powered applications across various domains – on the web, on mobile, and on edge devices. To support TensorFlow developers, there is comprehensive documentation and a large community of developers who report bugs and contribute to the further development and improvement of this framework.

Another central feature of the TensorFlow ecosystem is its access to a diverse array of datasets, cutting across different ML problem types such as image data, text data, and time-series data. These datasets are available via TensorFlow Datasets, and they are a great way to master the use of TensorFlow in solving real-world problems. In subsequent chapters, we will be exploring how to build models to solve computer vision, natural language processing, and time-series forecasting problems using a range of datasets available within the TensorFlow ecosystem.

We have explored some indispensable tools in the TensorFlow ecosystem. It is always a good idea to take a tour of these features (and the new ones that will be added) on the official website: `https://www.tensorflow.org/`. However, in the exam, you will not be quizzed on this. The idea here is to get familiar with what is available in the ecosystem. The exam focuses on modeling with TensorFlow so we will only use tools in the ecosystem such as TensorFlow Datasets, the Keras API, and TensorFlow Hub to meet this objective.

Setting up our environment

Before we examine data representations in TensorFlow, let's set up our work environment. We will begin by importing TensorFlow and checking the version:

```
import tensorflow as tf
#To check the version of TensorFlow
print(tf.__version__)
```

When we run this block of code, we get the following output:

```
2.8.0
```

Hurray! We have successfully imported TensorFlow. Next, let us import NumPy and a couple of data types, as we will be using them shortly in this chapter:

```
import numpy as np
from numpy import *
```

We have successfully completed all our import steps without errors. We will now look at data representations in TensorFlow as our working environment is fully set up.

Data representation

In our quest to solve complex tasks using ML, we come across diverse types of raw data. Our primary role involves transforming this raw data (which could be text, images, audio, or video) into numerical representations. These representations allow our ML models to easily digest and learn the underlying patterns in the data efficiently. To achieve this, this is where TensorFlow and its fundamental data structure, tensors, come into play. While numerical data is commonly used in training models, our models are also adept at efficiently handling binary and categorical data. For such data types, we apply techniques such as one-hot encoding to transform them into a model-friendly format.

Tensors are multi-dimensional arrays designed for numerical data representation; although they share some similarities with NumPy arrays, they possess certain unique features that give them an advantage in deep learning tasks. One of these key advantages is their ability to utilize hardware acceleration from GPUs and TPUs to significantly speed up computational operations, which is especially useful when working with input data such as images, text, and videos, as we will see in later chapters of this book.

Let us take a quick look at a real-world example. Let's say we are building an automobile recognition system, as illustrated in *Figure 2.1*. We would begin with collecting images of cars of various sizes, shapes, and colors. To train our model to recognize these different automobiles, we would transform each image into input tensors that encapsulate the height, width, and color channels. When we train the model on these input tensors, it learns patterns based on the pixel value representations of the cars in our training set. Once the model completes the training, we can use the trained model to identify cars of different shapes, colors, and sizes. If we now feed the trained model with the image of a car, it returns an output tensor that can be decoded into a human-readable format to enable us to identify the type of car that it is.

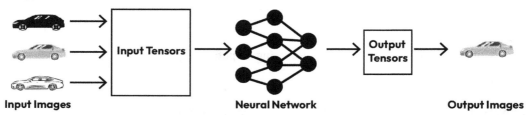

Figure 2.1 – Data representation in TensorFlow

Now that we get the intuition, let's examine and drill down into more details about tensors. We will start by learning a few ways to generate tensors next.

Creating tensors

There are a couple of ways we can generate tensors in TensorFlow. However, we will focus on creating tensor objects using tf.constant, tf.Variable, and tf.range. Recall that we have already imported TensorFlow, NumPy, and data types in the section on setting up our working environment. Next, let us run the following code to generate our first tensor using tf.constant:

```
#Creating a tensor object using tf.constant
a_constant = tf.constant([1, 2, 3, 4 ,5, 6])
a_constant
```

When we run this code, we generate our first tensor. If all goes well, we will get an output that looks like this:

```
<tf.Tensor: shape=(6,), dtype=int32,
    numpy=array([1, 2, 3, 4, 5, 6], dtype=int32)>
```

Excellent! Don't worry, we will discuss the output and form a clearer picture as we proceed. But for now, let us generate a similar tensor object using the tf.Variable function:

```
#Creating a tensor object using tf.Variable
a_variable = tf.Variable([1, 2, 3, 4 ,5, 6])
a_variable
```

The a_variable variable returns the following output:

```
<tf.Variable 'Variable:0' shape=(6,) dtype=int32,
    numpy=array([1, 2, 3, 4, 5, 6], dtype=int32)>
```

Although the input in both cases is the same, tf.constant and tf.Variable are different. Tensors generated using tf.constant cannot be changed, whereas tf.Variable tensors can be reassigned in the future. We will touch more on this shortly as we go further in our exploration of tensors. In the meantime, let us look at another way of generating tensors using tf.range:

```
# Creating tensors using the range function
a_range = tf.range(start=1, limit=7)
a_range
```

a_range returns the following output:

```
<tf.Tensor: shape=(6,), dtype=int32,
    numpy=array([1, 2, 3, 4, 5, 6], dtype=int32)>
```

Great! From the output, if we visually compare all three methods used for generating tensors, we can easily conclude that the output of a_constant and a_range is the same and is slightly different from the output of a_variable. This difference becomes clearer when performing tensor operations. To see this in action, let's begin exploring tensor operations, starting with tensor rank.

Tensor rank

If you are not from a mathematical background, relax. We will cover everything together and we won't be discussing rocket science here – that's a promise. The rank of a tensor identifies the number of dimensions of the tensor. A tensor with a rank of 0 is called a scalar, as it has no dimensions. A vector is a rank 1 tensor as it has only one dimension, while a matrix of a two-dimension tensor has a rank of 2.

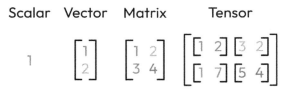

Figure 2.2 – Tensor rank

We have practiced how to generate tensors using three different functions. For context, we can safely define a scalar as a quantity that has only magnitude but no direction. Examples of scalar quantities are time, mass, energy, and speed; these quantities have a single numeric value, for example, 1, 23.4, or 50. Let us return to our notebook and generate a scalar using the tf.constant function:

```
#scalar
a = tf.constant(1)
a
```

We start by creating a scalar, which is a single value that returns the following output:

```
<tf.Tensor: shape=(), dtype=int32, numpy=1>
```

From the returned output, we can see that the shape has no value since the output is a scalar quantity with a single numeric output. If we try out a value of 4, the numpy output will be 4, while other output properties will remain the same since 4 is still a scalar quantity.

Now that we have seen what a scalar (rank 0 tensor) is, let us go a step further by looking at a vector. For context, a vector quantity has both magnitude and direction. Examples of vectors are acceleration, velocity, and force. Let us jump back into our notebook and try to generate a vector of four numbers. For a change, this time, we will use floats since we can generate tensors with floats. Also, if you noticed, the default data type returned has been int32 for integers, which we have previously used to generate tensors:

```
#vector
b= tf.constant([1.2,2.3,3.4,4.5])
b
```

From our result, we see the data type returned is float32 with a shape of 4:

```
<tf.Tensor: shape=(4,), dtype=float32,
    numpy=array([1.2, 2.3, 3.4, 4.5], dtype=float32)>
```

Next, let us generate a matrix. A matrix is an array of numbers listed in rows and columns. Let us try out a matrix in our notebook:

```
#matrix
c =tf.constant([[1,2],[3,4]])
c

<tf.Tensor: shape=(2, 2), dtype=int32,
    numpy= array([[1, 2], [3, 4]], dtype=int32)>
```

The preceding matrix is a 2 x 2 matrix, which we can infer by inspecting the shape output. We see that the data type is also int32. Let us generate a higher-dimensional tensor:

```
#3-dimensional tensor
d=tf.constant([[[1,2],[3,4],[5,6]],[[7,8],[9,10],[11,12]]])
d
```

The output is a 2 x 3 x 2 tensor, with a data type of int32:

```
<tf.Tensor: shape=(2, 3, 2), dtype=int32,
    numpy= array([[[ 1,    2],[ 3,    4],[ 5,    6]],
        [[ 7,    8], [ 9, 10], [11, 12]]], dtype=int32)>
```

You should play around with some tensors. Try making some tensors with tf.Variable and see whether you can reproduce our results so far. Next, let us see how we can interpret the properties of a tensor.

Properties of tensors

Now that we have established an understanding of scalars, vectors, and tensors, let us explore how to interpret tensor outputs in detail. Previously, we examined tensors with a piecemeal approach. Here, we will learn how to identify the key properties of a tensor – its rank, shape, and data type – from its printed representation. When we print a tensor, it displays the variable name, shape, and data type. Thus far, we utilized default arguments when creating tensors. Let us make some adjustments to see how this changes the output.

We will use tf.Variable to generate a scalar tensor, selecting float16 as the data type and naming it TDC. (If you are wondering what **TDC** means, it is the **TensorFlow Developer Certificate**.) Next, we will run the code:

```
#scalar
a = tf.Variable(1.1, name="TDC", dtype=float16)
a

<tf.Variable 'TDC:0' shape=() dtype=float16, numpy=1.1>
```

When we examine the output, we can see the name of the tensor is now TDC: 0, and the shape of the tensor is 0 since the tensor is of rank 0. The data type we selected was float16. And finally, the tensor has a numpy value of 1.1 also. This example shows how we can configure properties such as data type and name when constructing tensors in TensorFlow.

Next, let us look at a vector and see what information we can learn from its properties:

```
#vector
b= tf.Variable([1.2,2.3,3.4,4.5], name="Vector", dtype=float16)
b
```

Here, again, we included arguments and the name of the tensor and we changed the default data type. From the output, we can see the result is similar to what we got with the scalar quantity:

```
<tf.Variable ''Vector:0' shape=(4,) dtype=float16,
    numpy=array([1.2, 2.3, 3.4, 4.5])>
```

Here, the tensor has the name 'Vector:0, the shape has a value of 4 (which corresponds to the count of the number of entries), and the tensor has a data type of float16. To have some fun, you can experiment with different configurations and see the impact the changes you make have on the returned output; this is an excellent way to learn and understand how things work. When we print the result of a tensor output, we can see the different properties of the tensor, like when we examined the scalar and vector quantities. However, by leveraging TensorFlow functions, we can gain more information about a tensor. Let us start by using the tf.rank() function to inspect the rank of a scalar, vector, and matrix:

```
#scalar
a = tf.constant(1.1)
#vector
b= tf.constant([1.2,2.3,3.4,4.5])
#matrix
c =tf.constant([[1,2],[3,4]])
#Generating tensor rank
print("The rank of the scalar is: ",tf.rank(a))
print(" ")
print("The rank of the vector is: ",tf.rank(b))
print(" ")
print("The rank of the matrix is: ",tf.rank(c))
```

We run the preceding code to generate a scalar, vector, and matrix. After this, we print the rank of each of them using the tf.rank function. Here is the output:

```
The rank of the scalar is: tf.Tensor(0, shape=(), dtype=int32)
The rank of the vector is: tf.Tensor(1, shape=(), dtype=int32)
The rank of the matrix is: tf.Tensor(2, shape=(), dtype=int32)
```

The returned output is a tensor object that displays the rank of the tensors along with the shape and the data type of the tensor. To access the rank of the tensor as a numeric value, we have to use .numpy() on the returned tensor to retrieve the actual rank of the tensor:

```
print("The rank of the scalar is: ",tf.rank(a).numpy())
The rank of the scalar is:  0
```

However, an easier way to directly obtain the rank of a tensor without the need for reevaluating is by using ndim. Let's see this next:

```
#Generating details of the dimension

print("The dimension of the scalar is: ",a.ndim)
print(" ")
print("The dimension of the vector is: ",b.ndim)
print(" ")
print("The dimension of the matrix is: ",c.ndim)
```

When we run the code, we get the following output:

```
The dimension of the scalar is: 0
The dimension of the vector is: 1
The dimension of the matrix is: 2
```

Next, let us proceed by printing out the data type of all three quantities using the dtype argument to generate the data type for each of our tensors:

```
#printing the data type
print("The data type of the scalar is: ",a.dtype)
print(" ")
print("The data type of the vector is: ",b.dtype)
print(" ")
print("The data type of the matrix is: ",c.dtype)
```

When we run the code, we get the following output.

```
The data type of the scalar is:  <dtype: 'float32'>
The data type of the vector is:  <dtype: 'float32'>
The data type of the matrix is:  <dtype: 'int32'>
```

From the preceding output, we can see the data types. Next, let us look at the shape of our tensors:

```
#Generating details of the tensor shape
print("The Shape of the scalar is: ",a.shape)
print(" ")
print("The Shape of the vector is: ",b.shape)
```

```
print(" ")
print("The Shape of the matrix is: ",c.shape)
```

When we run the code, we get the following output:

```
The Shape of the scalar is:   ()
The Shape of the vector is:   (4,)
The Shape of the matrix is:   (2, 2)
```

From the results, we can see that the scalar has no shape value while the vector has a shape value of one unit, and our matrix has a shape value of two units. Next, let us compute the number of elements in each of our tensors:

```
#Generating number of elements in a tensor
print("The Size of the scalar is: ",tf.size(a))
print(" ")
print("The Size of the vector is: ",tf.size(b))
print(" ")
print("The Size of the matrix is: ",tf.size(c))
```

When we run the code, we get the following output:

```
The Size of the scalar is:   tf.Tensor(1, shape=(), dtype=int32)
The Size of the vector is:   tf.Tensor(4, shape=(), dtype=int32)
The Size of the matrix is:   tf.Tensor(4, shape=(), dtype=int32)
```

We can see that the scalar has only 1 count since it is a single unit; our vector and matrix both have 4 in them and hence they have 4 numeric values in each of them. Now, we can confidently use different ways to investigate the properties of tensors. Let us proceed to implement basic operations with tensors.

Basic tensor operations

We now know that TensorFlow is a powerful tool for deep learning. One big hurdle with learning TensorFlow is understanding what tensor operations are and why we need them. We have established that tensors are fundamental data structures in TensorFlow and they can be used to store, manipulate, and analyze data in ML models. On the other hand, tensor operations are mathematical operations that can be applied to tensors in order to manipulate, decode, or analyze data. These operations range from simple operations such as element-wise operations to more complex computations performed within the layers of a neural network. Let us look at some tensor operations. We will start with changing data types. Then, we will look at indexing and aggregating tensors. Finally, we will carry out element-wise operations on tensors, reshaping tensors, and matrix multiplication.

Changing data types

Let's say we have a tensor and we want to change the data type from int32 to float32, perhaps to accommodate some operation that would require the decimal numbers. Fortunately, in TensorFlow, there is a way around this problem. Remember that we identified that the default data type for integers is int32 and for decimal numbers, it is float32. Let us return to Google Colab and see how we can get this done in TensorFlow:

```
a=tf.constant([1,2,3,4,5])
a
```

We generated a vector of integers, which produces the following output:

```
<tf.Tensor: shape=(5,), dtype=int32, numpy=array([1, 2, 3, 4, 5],
dtype=int32)>
```

We can see that the data type is int32. Let us proceed with a data type operation, changing the data type to float32. We use the tf.cast() function and we set the data type argument to float32. Let us implement this in our notebook:

```
a =tf.cast(a,dtype=tf.float32)
a
```

The operation returns a data type of float32. We can also see the numpy array is now an array of decimal numbers and not integers anymore:

```
<tf.Tensor: shape=(5,), dtype=float32,
    numpy=array([1., 2., 3., 4., 5.], dtype=float32)>
```

You can try it out with int16 or float64 and see how it goes. When you are done, let us move on with indexing in TensorFlow.

Indexing

Let's start by creating a 2 x 2 matrix, which we will use to walk through our indexing operation:

```
# Create a 2 x 2 matrix
a = tf.constant([[1, 2],[3, 4]], dtype=float32)
a
```

Here is the returned output:

```
<tf.Tensor: shape=(2, 2), dtype=float32,
    numpy=array([[1., 2.], [3., 4.]], dtype=float32)>
```

What if we want to extract some information from the matrix? Let's say we want to extract [1, 2]. How do we go about this? Not to worry: we can apply indexing to get the desired information. Let us get it done in our notebook:

```
# Indexing
a[0]
```

Here is the returned output:

```
<tf.Tensor: shape=(2,), dtype=float32,
    numpy=array([1., 2.], dtype=float32)>
```

What if we want to extract value 2 from the matrix? Let us see how we can get it done:

```
# Indexing
a[0][1]
```

Here is the returned output:

```
<tf.Tensor: shape=(), dtype=float32, numpy=2.0>
```

Now, we have successfully extracted the value we wanted using indexing. To extract all the values in the matrix shown in *Figure 2.3*, we can use indexing to extract the desired element in the 2 x 2 matrix.

$$\begin{bmatrix} a[0][0] & a[0][1] \\ a[1][0] & a[1][1] \end{bmatrix}$$

Figure 2.3 – Matrix indexing

Next, let us look at another example of indexing – this time, using the tf.slice() function to extract information from a tensor:

```
c = tf.constant([0, 1, 2, 3, 4, 5])
print(tf.slice(c,begin=[2],size=[4]))
```

We generate a tensor, c. Then, we use the tf.slice function to slice the vector, starting at index 2 with a size or count of 4. When we run the code, we get the following result:

```
tf.Tensor([2 3 4 5], shape=(4,), dtype=int32)
```

We can see that the result contains values from index 2, and we take 4 elements in the vector to generate our slice. Next, let us look at how to expand the dimension of a matrix.

> **Important note**
> Remember, in Python, we start counting from 0, not 1.

Expanding a matrix

We already now know how to check the dimension of the matrix using ndim. So, let us see how we can expand the dimension of this matrix. We continue using our a matrix, which is a 2 x 2 matrix, as shown in *Figure 2.4*.

$$\begin{bmatrix} 1 & 2 \\ 3 & 4 \end{bmatrix}$$

Figure 2.4 – A 2x2 matrix

We can use the following code to expand the dimension:

```
tf.expand_dims(a,axis=0)
```

We use the expand_dims() function, and the code expands the dimensions of the a tensor along the 0 axis. This is useful when you want to add a new dimension to the tensor – for example, when you want to convert a 2D tensor into a 3D tensor (a technique that will be applied in *Chapter 7*, *Image Classification with Convolutional Neural Networks*, where we will work on an interesting classic image dataset):

```
<tf.Tensor: shape=(1, 2, 2), dtype=float32,
    numpy= array([[[1., 2.], [3., 4.]]], dtype=float32)>
```

If you take a look at the shape of our output tensor, we can now see it has an extra dimension of 1 at the 0 axis. Let us proceed by examining the shape of the tensor when we expand across different axes, so we can understand how this works better:

```
(tf.expand_dims(a,axis=0)).shape,
(tf.expand_dims(a,axis=1)).shape,
(tf.expand_dims(a,axis=-1)).shape
```

When we run the code to see how the dimension has expanded across the 0, 1, and -1 axes, we get the following results:

```
(TensorShape([1, 2, 2]), TensorShape([2, 1, 2]),
    TensorShape([2, 2, 1]))
```

In the first line of code, the dimensions of a are expanded by 1 on the 0 axis. This means that the dimensions of a will now be 1 x 2 x 2, adding an extra dimension at the beginning of the tensor. The second line of code is expanding the dimensions of a by 1 on the 1 axis. This means that the dimensions of a will now be 2 x 1 x 2; here, we are adding an extra dimension in the second position of the tensor. The third line of code is expanding the dimensions of a by 1 on the -1 axis. This means that the dimensions of a will now be 2 x 2 x 1, thereby adding an extra dimension at the end of our tensor. We have now explained how to expand the dimension of a matrix. Next, let us look at tensor aggregation.

Tensor aggregation

Let us continue our journey by understanding how to aggregate tensors. We start by generating some random numbers by importing the random library. Then, we generate a range from 1 to 100 in which we generate 50 random numbers. We will now use these random numbers to generate a tensor:

```
import random
random.seed(22)
a = random.sample(range(1, 100), 50)
a = tf.constant(a)
```

When we print a, we get the following numbers:

```
<tf.Tensor: shape=(50,), dtype=int32, numpy=array(
    [16, 83,  6, 74, 19, 80, 95, 68, 66, 86, 54, 12, 91,
    13, 23,  9, 82, 84, 30, 62, 89, 33, 78,  2, 97, 21,
    59, 34, 48, 38, 35, 18, 46, 60, 27, 26, 73, 76, 94,
    72, 15, 40, 96, 44, 61,  8, 79, 93, 11, 14],
    dtype=int32) >
```

Let's say we want to find the smallest number in our tensor. It may be difficult to manually read through all the numbers and tell me in 5 seconds which is the smallest. What if our range of values was up to a thousand or a million? Manually checking would take up all our time. Thankfully, in TensorFlow, we can find not just the minimum in one strike but we can also find the maximum value, the sum of all values, the mean, and much more. Let us do this together in the Colab notebook:

```
print("The smallest number in our vector is : ",
    tf.reduce_min(a).numpy())
print(" ")
print("The largest number in our vector is: ",
    tf.reduce_max(a).numpy())
print(" ")
print("The sum of our vector is : ",
    tf.reduce_sum(a).numpy())
print(" ")
print("The mean of our vector is: ",
    tf.reduce_mean(a).numpy())
```

We use these functions to extract the details we require in one click, which generates the following result:

```
The smallest number in our vector is :  1

The largest number in our vector is:   99

The sum of our vector is :   2273

The mean of our vector is:   45
```

Now that we have used TensorFlow to extract some important details, we know that the smallest value in our vector is 1, the largest value is 99, the sum of our vector is 2,273, and the mean value is 45. Not bad, right? What if we want to find the position that holds the minimum and maximum value in a vector? How do we go about this?

```
print("The position that holds the lowest value is : ",
    tf.argmin(a).numpy())
print(" ")
print("The position that holds the highest value is: ",
    tf.argmax(a).numpy())
```

We use the `tf.argmin` and `tf.argmax` functions to generate the index of the lowest value and the index of the highest value, respectively. The output is as follows:

```
The position that holds the lowest value is :  14
The position that holds the highest value is:  44
```

From the result of the `print` statement, we can tell that the lowest value is at index 14 and the highest value is at index 44. If we manually inspect the array, we will see that this is true. Also, we can pass the index position into the array to get the lowest and highest value:

```
a[14].numpy(), a[44].numpy()
```

If we run the code, we get the following result:

```
(1,99)
```

There are a few other functions you can try out. The TensorFlow documentation gives us a lot to try out and have fun with. Next, let us explore how to transpose and reshape tensors.

Transposing and reshaping tensors

Let us look at how to transpose and reshape a matrix. First, let's generate a 3 x 4 matrix:

```
# Create a 3 x 4 matrix
a = tf.constant([[1,2,3,4], [5,6,7,8], [9,10,11,12]])
a
```

When we run the code, we get this result:

```
<tf.Tensor: shape=(3, 4), dtype=int32,
    numpy=array([[ 1,   2,   3,   4],
         [ 5,   6,   7,   8],
         [ 9, 10, 11, 12]], dtype=int32)>
```

We can reshape the matrix by using `tf.reshape`. Since the matrix has 12 values in it, we can use 2 x 2 x 3. If we multiply the values, we get a total of 12:

```
tf.reshape(a, shape=(2, 2, 3))
```

When we run the code, we get the following output:

```
<tf.Tensor: shape=(2, 2, 3), dtype=int32,
    numpy=array([[[ 1,   2,   3], [ 4,   5,   6]],
         [[ 7,   8,   9], [10, 11, 12]]], dtype=int32)>
```

We can also reshape the matrix by changing the `shape` argument in the `tf.reshape` function to a 4 x 3 matrix or a 1 x 2 x 6 matrix. You can also try out a few other possibilities with regard to reshaping this matrix. Next, let us look at how to transpose this matrix using `tf.transpose()`:

```
tf.transpose(a)
```

When we run the code, we get the following output:

```
<tf.Tensor: shape=(4, 3), dtype=int32,
    numpy=array([[ 1,   5,   9],
         [ 2,   6, 10],
         [ 3,   7, 11],
         [ 4,   8, 12]], dtype=int32)>
```

From the output, we can see that transposing flips the axes. We now have a 4 x 3 matrix rather than our initial 3 x 4 matrices. Next, let us look at element-wise matrix operations.

Element-wise operations

Let's start by creating a simple vector in Colab:

```
a= tf.constant([1,2,3])
a
```

Let us display our output so we can see what happens when we perform element-wise operations on the vector:

```
<tf.Tensor: shape=(3,), dtype=int32, numpy=array([1, 2, 3],
    dtype=int32)>
```

This is our initial output. Now, let us try out a few element-wise operations and see what happens next:

```
#Addition operation
print((a+4).numpy())
print(" ")
#Subtraction operation
print((a-4).numpy())
print(" ")
#Multiplication Operation
print((a*4).numpy())
print(" ")
#Division Operation
print((a/4).numpy())
print(" ")
```

We can see the results for the addition, subtraction, multiplication, and division operations. These operations are carried out on each element in our vector:

```
[5  6  7]
[-3  -2  -1]
[ 4   8  12]
[0.25 0.5  0.75]
```

Next, let us look at matrix multiplication.

Matrix multiplication

Let us look at matrix multiplication and see how it works in TensorFlow. We return to our notebook in Colab and generate matrix a, which is a 3 x 2 matrix, and matrix b, which is a 2 x 3 matrix. We will use these for our matrix operations:

```
# 3 X 2 MATRIX
a = tf.constant([[1, 2], [3, 4], [5, 6]])

#2 X 3 MATRIX
b = tf.constant([[7,8,9], [10,11,12]])
```

Now, let us multiply matrix a and b and see what our result will look like in TensorFlow by using tf.matmul in our notebook:

```
tf.matmul(a,b)
```

We use the `tf.matmul` function for matrix multiplication in TensorFlow. Here, we see the output of this operation:

```
<tf.Tensor: shape=(3, 3), dtype=int32,
 numpy= array([[ 27,   30,   33],[ 61,   68,   75],
    [ 95, 106, 117]], dtype=int32)>
```

Great! Now, what if we want to multiply matrix a by itself? What will our result look like? If we tried this out, we will get an error because the shape of the matrix does not conform to the rule of matrix multiplication. The rule requires that matrix a should be made up of *i* rows x *m* columns, and matrix b should be made up of *m* rows x *n* columns, where the value of *m* must be the same in both matrices. The new matrix will have a shape of *i* x *n*, as shown in *Figure 2.5*.

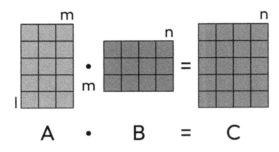

Figure 2.5 – Matrix multiplication

Now we can see why we cannot multiply matrix a by itself, because the number of rows in the first matrix must be equal to the number of columns in the second matrix. However, we can fulfill the requirement of the matrix multiplication rule by either transposing or reshaping matric a if we want to multiply a by itself. Let us try this out:

```
tf.matmul(a,tf.transpose(a, perm=[1,0]))
```

When we transpose matrix a, we swap the rows and columns of the matrix based on the `perm` parameter, which we set to `[1,0]`. When we execute the `matmul` function using a and the transpose of a, we get a new matrix that complies with the rule of matrix multiplication:

```
<tf.Tensor: shape=(3, 3), dtype=int32,
    numpy= array([[ 5, 11, 17], [11, 25, 39],
    [17, 39, 61]], dtype=int32)>
```

How about we try out matrix multiplication using `reshape`? Give this a shot and compare your result with our working Colab notebook results. We have looked at a whole lot of operations already. How about we build our first model? Let us do that next.

Hello World in TensorFlow

We have covered a lot of basic operations in TensorFlow. Now, let's build our first model in TensorFlow. For this example, let us say you are part of a research team studying the correlation between the number of hours a student studied in a term and their final grade. Of course, this is a theoretical scenario and there are a lot more factors that come into play when it comes to how well a student will perform. However, in this case, we will take only one attribute as the determinant of success – hours of study. After a term of study, we successfully collated the hours of study of students and their corresponding grades, as shown in *Table 2.1*.

Hours of Study	20	23	25	28	30	37	40	43	46
Test Score	45	51	55	61	65	79	85	91	97

Table 2.1 – Students' performance table

Now, we want to build a model to predict how well a student will perform in the future based on the hours of study they put in. Ready? Let's do this together now:

1. Let's build this together by opening the accompanying notebook called `hello world`. First, we import TensorFlow. Remember in *Chapter 1*, *Introduction to Machine Learning*, we talked about features and labels. Here, we just have one feature – hours of study – and our label or target variable is the test score. Using the powerful Keras API, in a few lines of code, we will build and train a model to get predictions. Let's get started:

    ```
    import tensorflow as tf
    from tensorflow import keras
    from tensorflow.keras import Sequential
    from tensorflow.keras.layers import Dense
    print(tf.__version__)
    ```

    We start by importing TensorFlow and the Keras API; don't worry about all the terms, we will unpack everything in detail in *Chapter 3*, *Linear Regression with TensorFlow*. The goal here is to show you how we build a basic model. Don't worry so much about the technicalities; running the code and seeing how it works is the goal here. After importing the necessary libraries, we continue with our tradition of printing our TensorFlow version. The code runs fine.

2. Next, we proceed to import `numpy` for carrying out mathematical operations and `matplotlib` for visualizing our data:

    ```
    #import additional libraries
    import numpy as np
    import matplotlib.pyplot as plt
    ```

    We run the code and we get no errors, so we are good to proceed.

3. We set up a list of X and y values representing our hours of study and test scores, respectively:

```
# Hours of study
X = [20,23,25,28,30,37,40,43,46]
# Test Scores
y = [45, 51, 55, 61, 65, 79, 85, 91, 97]
```

4. To get a good sense of data distribution, we use `matplotlib` to visualize our data:

```
plt.plot(X, y)
plt.title("Exam Performance graph")
plt.xlabel('Hours of Study')
plt.ylabel('Test Score')
plt.show()
```

The code block plots a graph of X (hours of study) against y (test score) and displays the title (Exam Performance graph) of our plot. We use the show() function to display the graph, as shown in *Figure 2.6*.

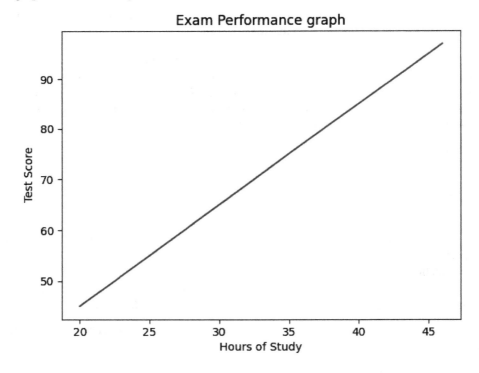

Figure 2.6 – Students' performance plot

From the plot, we can see the data shows a linear relationship. This assumption is not so bad considering we would logically expect a student who works harder to score better marks.

5. Without getting into a debate about whether this theory holds, let us use the Keras API to build a simple model:

```
study_model = Sequential([Dense(units=1,
    input_shape=[1])])
study_model.compile(optimizer='adam',
    loss='mean_squared_error')
X= np.array(X, dtype=int)
y= np.array(y, dtype=int)
```

We build a one-layer model, which we call `study_model`, and we convert our list of X and y values into a NumPy array.

6. Next, we fit our model and run it for 2,500 epochs:

```
#fitting the model
history= study_model.fit(X, y, epochs=2500)
```

When we run the model, it should take less than 5 minutes. We can see that the loss drops rapidly initially and gradually flattens out at around 2,000, epochs as shown in *Figure 2.8*:

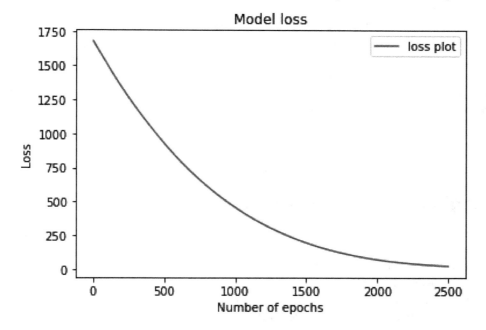

Figure 2.7 – Model loss plot

And just like that, we have trained a model that can be used to determine how a student will perform at the end of a term. This is a very basic task and feels like using a hammer on a fly. However, let us try our model out:

```
#Let us predict how well a student will perform based on their study
time
n=38 #Hours of study
result =study_model.predict([n])[0][0] #Result
rounded_number = round(81.0729751586914, 2)
```

If we run this code, we generate the result for a student who studied for 38 hours. Remember our model was not trained on this value. So, let us see what our model thinks this student will score:

```
print(f"If I study for {n} hours,
    I will get { rounded_number} marks as my grade.")
If I study for 38 hours, I will get 81.07 marks as my grade.
```

Our model predicted that this student would score 81.07 marks. Good result, but how do we know whether our model was right or wrong? If you look at *Figure 2.6*, you may guess that our predicted result should be around this score, but you may also have figured out that we used *2x + 5 = y* to generate our y values. If we input X=38, we get *2(38) + 5= 81*. Our model did an excellent job of getting the correct score with a minute error of .07; however, we had to train it for a very long time to achieve this result for a very simple task. In the coming chapters, we will learn how to train a much better model using techniques such as normalization and, of course, with a larger dataset, where we will work with a training set and a validation set and make predictions on a test set. The goal here was to get a feel of what is to come, so try out a few numbers to see how the model will perform. Do not go above 47, as you will get a score above 100.

Now that we have built our first model, let us look at how to debug and solve error messages. This is something you will encounter many times if you decide to pursue a career in this space.

Debugging and solving error messages

As you go through the exercises or walk through the code in this book, in any other resource, or in your own personal projects, you will quickly realize how often code breaks, and mastering how to resolve these errors will help you to move quickly through your learning process or when building projects. First, when you get an error, it is important to check what the error message is. Next is to understand the meaning of the error message. Let us look at some errors that a few students stumbled upon when implementing basic operations in TensorFlow. Let's run the following code to generate a new vector:

```
tf.variable([1,2,3,4])
```

Running this code will throw the error shown in the following screenshot:

```
- - - - - - - - - - - - - - - - - - - - - - - - - - - - - - - - - - - - - - - - - - -
AttributeError                              Traceback (most recent call last)
<ipython-input-123-c5af75a75660> in <module>()
----> 1 tf.variable([1,2,3,4])
```

AttributeError: module 'tensorflow' has no attribute 'variable'

SEARCH STACK OVERFLOW

Figure 2.8 – Example of an error

From the error message, we can see that there is no attribute called `variable` in TensorFlow. This draws our attention to where the error is coming from and we immediately notice that we wrote `variable` instead of `Variable` with a capital *V*, as stipulated in the documentation. However, if we are not able to debug this ourselves, we can click on the **SEARCH STACK OVERFLOW** button, as this is a good place to find solutions to everyday coding problems we might encounter. The odds are someone else has faced the same problem and a solution can be found on Stack Overflow.

Let us click on the link and see what we can find on Stack Overflow:

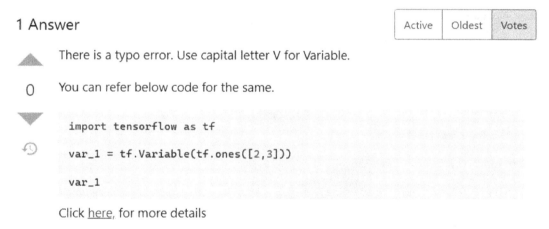

Figure 2.9 – Stack Overflow solution for AttributeError

Hurray! On Stack Overflow, we see the solution to the problem and a link to the documentation for more details. Remember, it is best to first look at the error message and see whether you can resolve it yourself before heading to Stack Overflow. If you put this into practice, as well as dedicate time to reading the documentation, you will get better and better at debugging issues and make fewer mistakes, but you will still need Stack Overflow or the documentation. It comes with the terrain. Before we draw the curtains on this chapter, let us summarize quickly what we learned.

Summary

In this chapter, we covered the TensorFlow ecosystem at a high level. We looked at some of the key components that make TensorFlow the platform of choice for building deep learning applications and solutions for many ML engineers, researchers, and enthusiasts. Next, we discussed what tensors are and how they are useful in our models. After this, we looked at a few ways of creating tensors. We explored various tensor properties and we saw how to implement some basic tensor operations with TensorFlow. We built a simple model and used it to make predictions. Finally, we looked at how to debug and solve error messages in TensorFlow and ML at large.

In the next chapter, we will look at regression modeling in a hands-on manner. We will learn how to extend our simple model to solve a regression problem for a company's HR department. Also, what you have learned about debugging could prove useful in the next chapter – see you there.

Questions

Let's test what we have learned in this chapter:

1. What is TensorFlow?
2. What are tensors?
3. Generate a matrix using `tf.Variable` with the `tf.float64` data type and name the variable.
4. Generate 15 random numbers between 1 and 20 and extract the lowest number, the highest number, the mean, and the index with the lowest and highest numbers.
5. Generate a 4 x 3 matrix, and multiply the matrix by its transpose.

Further reading

To learn more, you can check out the following resources:

- Amr, T., 2020. *Hands-On Machine Learning with scikit-learn and Scientific Python Toolkits.* [S.l.]: Packt Publishing.

- *TensorFlow guide*: `https://www.TensorFlow.org/guide`

3

Linear Regression
with TensorFlow

In this chapter, we will cover the concept of linear regression and how we can implement it using TensorFlow. We will start by discussing what linear regression is, how it works, its underlying assumptions, and the type of problems that can be solved using it. Next, we will examine the various evaluation metrics used in regression modeling, such as mean squared error, mean absolute error, root mean squared error, and R-squared, and strive to understand how to interpret the results from these metrics.

To get hands-on, we will implement linear regression by building a real-world use case where we predict employees' salaries using various attributes. Here, we will learn in a hands-on fashion how to load and pre-process data, covering important ideas such as handling missing values, encoding categorical variables, and normalizing the data for modeling. Then, we will explore the process of building, compiling, and fitting a linear regression model with TensorFlow, as well as examine concepts such as underfitting and overfitting and their impact on our model's performance. Before we close this chapter, you will also learn how to save and load a trained model to make predictions on unseen data.

In this chapter, we'll cover the following topics:

- Linear regression with TensorFlow
- Evaluating regression models
- Salary prediction with TensorFlow
- Saving and loading models

Technical requirements

We will use a **Google Colaboratory** (**Google Colab**) notebook as our work environment, as it is a free, cloud-based Jupyter notebook that is easy to use, providing us with GPU and TPU backends. For the coding exercises in this chapter, we will use `python >= 3.8.0`, along with the following packages, which can be installed using the `pip install` command:

- `tensorflow>=2.7.0`
- `tensorflow-datasets==4.4.0`
- `pillow==8.4.0`
- `pandas==1.3.4`
- `numpy==1.21.4`
- `scipy==1.7.3`

Linear regression with TensorFlow

Linear regression is a supervised machine learning technique that models the linear relationship between the predicted output variable (dependent variable) and one or more independent variables. When one independent variable can be used to effectively predict the output variable, we have a case of *simple linear regression*, which can be represented by the equation $y = wX + b$, where y is the target variable, X is the input variable, w is the weight of the feature(s), and b is the bias.

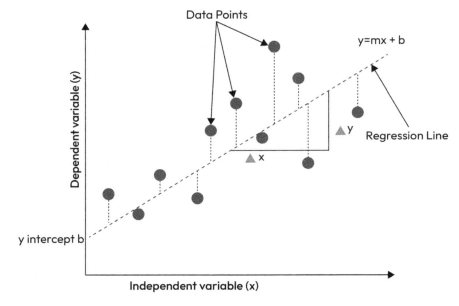

Figure 3.1 – A plot showing simple linear regression

In *Figure 3.1*, the straight line, referred to as the regression line (the line of best fit), is the line that optimally models the relationship between X and y. Hence, we can use it to determine the dependent variable based on the current value of the independent variable at a certain point on the plot. The objective of linear regression is to find the best values of w and b, which model the underlying relationship between X and y. The closer the predicted value is to the ground truth, the smaller the error.

Conversely, when we have more than one input variable used to predict the output value, then we have a case of *multiple linear regression*, and we can represent it by the equation $y = b0 + b1X1 + b2X2 + ... + bnXn$, where y is the target variable, $X1$, $X2$, ... Xn are input variables, $b0$ is the bias, and $b1$, $b2$, ... bn are the feature weights.

Simple linear and multiple linear regression have lots of real-world applications, as they are simple to implement and computationally cheap. Hence, they can be easily applied to large datasets. However, linear regression may fail when we try to model nonlinear relationships between X and y, or when there are many irrelevant features in our input data.

Linear regression is widely used to solve a wide range of real-world problems across different domains. For example, we can apply linear regression to predict the price of a house using factors such as the size, number of bedrooms, location, and proximity to social amenities. Similarly, in the field of **human resource (HR)**, we can use linear regression to predict the salary of new hires, based on factors such as years of experience of the candidate and their level of education. These are a few examples of what is possible using linear regression. Next, let us see how we can evaluate linear regression models.

Evaluating regression models

In our *hello world* example from *Chapter 2, Introduction to TensorFlow*, we tried to predict a student's test score when the student spent 38 hours studying during the term. Our study model arrived at 81.07 marks, while the true value was 81. So, we were close but not completely correct. When we subtract the difference between our model's prediction and the ground truth, we get a residual of 0.07. The residual value could be either positive or negative, depending on whether our model overestimates or underestimates the predicted result. When we take the absolute value of the residual, we eliminate any negative signs; hence, the absolute error will always be a positive value, irrespective of whether the residual is positive or negative.

The formula for absolute error is as follows:

$$Absolute\ error = \left| Y_{pred} - Y_{true} \right|$$

where Y_{pred} = the predicted value and Y_{true} = the ground truth.

The **mean absolute error (MAE)** of a model is the average of all absolute errors of the data points under consideration. MAE measures the average of the residuals and can be represented using the following equation:

$$MAE = \frac{1}{n} \sum_{i=1}^{n} \left| Y_{pred} - Y_{true} \right|$$

where:

- n = the number of data points under consideration

- Σ = summation of the absolute errors of all the observations

- $\left| Y_{pred} - Y_{true} \right|$ = Absolute value

If the MAE = 0, it means that $Y_{pred} = Y_{true}$. This means the model is 100 percent accurate; although this is an ideal scenario, it is highly unlikely. On the flip side, if MAE= ∞, this means the model is completely off, as it fails to capture any relationship between the input and output variables. The larger the error, the larger the value of the MAE. For performance evaluation, we aim for low values of MAE, but because MAE is a relative metric whose value depends on the scale of the data you work with, it is difficult to compare MAE results across different datasets.

Another important evaluation metric is the **mean squared error** (**MSE**). MSE, in contrast to MAE, squares the residuals, thus removing any negative values in the residuals. MSE is represented using the following equation:

$$MSE = \frac{1}{N} \sum_{i=1}^{N} \left(Y_{pred} - Y_{true} \right)^2$$

Like MAE, when there are no residuals, we have a perfect model. So, the lower the MSE value, the better the performance of the model. Unlike MAE, where large or small errors have a proportional impact, MSE penalizes larger errors in comparison to smaller errors, and it has a higher order of units, since we square the residual in this instant.

Another useful metric in regression modeling is the **root mean square error** (**RMSE**). As the name suggests, it is the square root of MSE, as shown in the equation:

$$MSE = \frac{1}{N} \sum_{i=1}^{N} \left(Y_{pred} - Y_{true} \right)^2$$

$$RMSE = \sqrt{MSE} = \sqrt{\frac{1}{N} \sum_{i=1}^{N} \left(Y_{pred} - Y_{true} \right)^2}$$

Lastly, let us look at the **coefficient of determination** (**R squared**). R^2 measures how well the dependent variable is explained by the independent variables in a regression modeling task. We can calculate R^2 with this equation:

$$R^2 = 1 - \frac{R_{res}}{R_{tot}}$$

where R_{res} is the sum of the square of residuals and R_{tot} is the total sum of squares. The closer the value of R^2 is to 1, the more accurate the model is, and the closer the R^2 value of a model is to 0, the worse the model is. Also, it is possible for R^2 to take on a negative value. This happens when the model does not follow the trend of the data – in this instance, R_{res} is greater than R_{tot}. A negative R^2 is a sign that our model requires significant improvement due to its poor performance.

We have looked at some regression evaluation metrics. The good news is that we will not work them out by hand; we will leverage the `tf.keras.metrics` module from TensorFlow to help us do the heavy lifting. We have breezed quickly through the theory at a high level. Now, let us examine a multiple linear regression case study to enable us to understand all the moving parts required to build a model with TensorFlow, as well as understand how to evaluate, save, load, and use our trained model to make predictions on new data. Let's proceed to our case study.

Salary prediction with TensorFlow

In this case study, you will assume the role of a new machine learning engineer at Tensor Limited, a rapidly growing start-up with over 200 employees. Now, the company wants to hire seven new employees, and the HR department is having a hard time coming up with the ideal salary based on varying qualifications, years of experience, the roles applied for, and the level of training of each of the potential new hires. Your job is to work with the HR unit to determine the optimal salary for each of these potential hires.

Luckily, we went through the machine learning life cycle in *Chapter 1, Introduction to Machine Learning,* built our hello world case study in *Chapter 2, Introduction to TensorFlow,* and have already covered some key evaluation metrics required for regression modeling in this chapter. So, you are well equipped theoretically to carry out the task. You have had a productive discussion with the HR manager, and now you have a better understanding of the task and the requirements. You defined your task as a supervised learning task (regression). Also, the HR unit allowed you to download employee records and their corresponding salaries for this task. Now that you have the dataset, let us proceed to load the data into our notebook.

Loading the data

Perform the following steps to load the dataset:

1. Open the notebook called `Linear_Regression_with_TensorFlow.ipynb`. We will start by importing all the necessary libraries for this project:

    ```
    # import tensorflow
    import tensorflow as tf
    from tensorflow import keras
    from tensorflow.keras import Sequential
    from tensorflow.keras.layers import Dense
    print(tf.__version__)
    ```

    We will run this code block. If everything goes well, we will get to see the version of TensorFlow we are using:

    ```
    2.12.0
    ```

2. Next, we will import some additional libraries that will help us simplify our workflow:

```
import numpy as np
import pandas as pd
import matplotlib.pyplot as plt
import seaborn as sns
from sklearn.model_selection import train_test_split
from sklearn.preprocessing import MinMaxScaler
```

We will run this cell, and everything should work perfectly. NumPy is a scientific computing library in Python that is used to perform mathematical operations on arrays, while pandas is a built-in Python library for data analysis and manipulation. Matplotlib and Seaborn are used to visualize data, and we will use sklearn for data preprocessing and splitting our data. We will apply these libraries in this case study, and you will get to understand what they do and also be able to apply them in your exam and beyond.

3. Now, we will proceed to load the dataset, which we got from the HR team for this project:

```
#Loading from the course GitHub account
df=pd.read_csv('https://raw.githubusercontent.com/oluwole-packt/
datasets/main/salary_dataset.csv')
df.head()
```

We will use pandas to generate a DataFrame that holds the record in a tabular format, and we will use `df.head()` to print the first five entries in the data:

	Name	Phone_Number	Experience	Qualification	University	Role	Cert	Date_Of_Birth	Salary
0	Jennifer Hernandez	120-602-1220	3.0	Msc	Tier2	Mid	Yes	25/08/1972	98000
1	Timothy Walker	840-675-8650	5.0	PhD	Tier2	Senior	Yes	03/12/2013	135500
2	David Duran	556-293-8643	5.0	Msc	Tier2	Senior	Yes	19/07/2002	123500
3	Gloria Ortega	463-559-7474	3.0	Bsc	Tier3	Mid	No	19/02/1970	85000
4	Matthew Steele	968-091-7683	5.0	Bsc	Tier2	Senior	Yes	20/02/1970	111500

Figure 3.2 – A DataFrame showing a snapshot of our dataset

We now have a sense of what data was collected, based on the details captured in each column. We will proceed to explore the data to see what we can learn and how we can effectively develop a solution to meet the business objective. Let us proceed by looking at data pre-processing.

Data preprocessing

To be able to model our data, we need to ensure it is in the right form (i.e., numerical values). Also, we will need to deal with missing values and remove irrelevant features. In the real world, data preprocessing takes a long time. You will hear this repeatedly, and it is true. Without correctly shaping the data, we cannot model it. Let's jump in and see how we can do this for our current task. From the DataFrame, we can immediately see that there are some irrelevant columns, and they hold personally identifiable information about employees. So, we will remove and also inform HR about this:

```
#drop irrelevant columns
df =df.drop(columns =['Name', 'Phone_Number',
    'Date_Of_Birth'])
df.head()
```

We will use the `drop` function in pandas to drop the name, phone number, and date of birth columns. We will now display the DataFrame again using `df.head()` to show the first five rows of the data:

	Experience	Qualification	University	Role	Cert	Salary
0	3.0	Msc	Tier2	Mid	Yes	98000
1	5.0	PhD	Tier2	Senior	Yes	135500
2	5.0	Msc	Tier2	Senior	Yes	123500
3	3.0	Bsc	Tier3	Mid	No	85000
4	5.0	Bsc	Tier2	Senior	Yes	111500

Figure 3.3 – The first five rows of the DataFrame after dropping the columns

We have successfully removed the irrelevant columns, so we can now proceed and check for missing values in our dataset using the `isnull()` function in pandas:

```
#check the data for any missing values
df.isnull().sum()
```

When we run this code block, we can see that there are no missing values in the University and Salary columns. However, we have missing values for the Role, Cert, Qualification, and Experience columns:

```
Experience       2
Qualification    1
University       0
Role             3
Cert             2
Salary           0
dtype: int64
```

There are a number of ways to handle missing values – from simply asking HR to fix the omissions to simple imputations or replacements using mean, median, or mode. In this case study, we will drop the rows with missing values, since it's a small subset of our data:

```
#drop the null values
df=df.dropna()
```

We use the `dropna` function to drop all the missing values in the dataset, and then we save the new dataset in `df`.

> **Note**
>
> If you want to learn more about how to handle missing values, check out this playlist by Data Scholar: `https://www.youtube.com/playlist?list=PLB9iiBW-oO9eMF45oEMB5pvC7fsqgQv7u`.

Now, we need to check to ensure that there are no more missing values using the `isnull()` function:

```
#check for null values
df.isnull().sum()
```

Run the code, and let's see whether there are any missing values:

```
Experience     0
Qualification  0
University     0
Role           0
Cert           0
Salary         0
dtype: int64
```

We can see that there are no missing values in our dataset anymore. Our model requires us to pass in numerical values for it to be able to model our data and predict the target variable, so let us look at the data types:

```
df.dtypes
```

When we run the code, we get an output showing the different columns and their data types:

```
Experience     float64
Qualification   object
University      object
Role            object
Cert            object
Salary           int64
dtype: object
```

From the output, we can see that experience and salary are numeric values, since they are `float` and `int`, respectively, while `Qualification`, `University`, `Role`, and `Cert` are categorical values. This means we cannot train our model yet; we have to find a way to convert our categorical values to numerical values. Luckily, this is possible via a process called one-hot encoding. **One-hot**

encoding is a technique in which we convert the categorical variables in our data to individual columns representing each of the categories. We will use the `get_dummies` function in pandas to achieve this:

```
#Converting categorical variables to numeric values
df = pd.get_dummies(df, drop_first=True)
df.head()
```

When we run the code, we will get a DataFrame like the one displayed in *Figure 3.4*. We use the `drop_first` argument to drop the first category.

	Experience	Salary	Qualification_Msc	Qualification_PhD	University_Tier2	University_Tier3	Role_Mid	Role_Senior	Cert_Yes
0	3.0	98000	1	0	1	0	1	0	1
1	5.0	135500	0	1	1	0	0	1	1
2	5.0	123500	1	0	1	0	0	1	1
3	3.0	85000	0	0	0	1	1	0	0
4	5.0	111500	0	0	1	0	0	1	1

Figure 3.4 – A DataFrame showing numerical values

If you are confused as to why we dropped one of the categorical columns, let's look at the Cert column, which was made up of yes or no. values If we performed one hot encoding, without dropping any columns, we would have two Cert columns, as displayed in *Figure 3.5*. In the Cert_No column, if the employee has a relevant certification, the column gets a value of 0, and when the employee does not have a relevant certification, the column gets a value of 1. Looking at the Cert_Yes column, we can see that when an employee has a certificate, the column gets a value of 1; otherwise, it gets 0.

Cert_No	Cert_Yes
0	1
0	1
0	1
1	0
0	1

Figure 3.5 – The dummy variables from the Cert column

From *Figure 3.5*, we can see that both columns can be used to show whether an employee has a certificate or not. Using both dummy columns generated from our certificate column will lead to the *dummy variable trap*. This occurs when our one-hot encoded columns are strongly related or correlated, where one column can effectively explain the other column. Hence, we say both columns are multicollinear, and *multicollinearity* can lead to the overfitting of our model. We will talk more about overfitting in *Chapter 5, Image Classification with Neural Networks*.

For now, it is good enough to know that overfitting is a situation where our model performs very well on training data but poorly on test data. To avoid the dummy variable trap, we will drop one of the columns in *Figure 3.5*. If there are three categories, we only need two columns to capture all three categories; if we have four categories, we will only need three columns to capture four categories, and so on. Hence, we can drop the extra columns for all the other categorical columns as well.

Now, we will use the `corr()` function to get the correlation of our refined dataset:

```
df.corr()
```

We can see that there is a strong correlation between salary and years of experience. Also, there is a strong correlation between `Role_Senior` and `Salary`, as shown in *Figure 3.6*.

	Experience	Salary	Qualification_Msc	Qualification_PhD	University_Tier2	University_Tier3	Role_Mid	Role_Senior	Cert_Yes
Experience	1.000000	0.814831	0.120081	-0.044225	0.017854	-0.010583	-0.353572	0.889459	-0.002359
Salary	0.814831	1.000000	0.118207	0.365999	0.112023	0.057214	-0.156297	0.822280	0.150998
Qualification_Msc	0.120081	0.118207	1.000000	-0.363826	0.116238	-0.245785	-0.095579	0.112477	-0.033259
Qualification_PhD	-0.044225	0.365999	-0.363826	1.000000	0.142149	-0.054767	0.129437	-0.021039	-0.001612
University_Tier2	0.017854	0.112023	0.116238	0.142149	1.000000	-0.448447	-0.013210	0.030849	-0.003412
University_Tier3	-0.010583	0.057214	-0.245785	-0.054767	-0.448447	1.000000	-0.022069	-0.002311	0.016894
Role_Mid	-0.353572	-0.156297	-0.095579	0.129437	-0.013210	-0.022069	1.000000	-0.544499	0.020775
Role_Senior	0.889459	0.822280	0.112477	-0.021039	0.030849	-0.002311	-0.544499	1.000000	0.002759
Cert_Yes	-0.002359	0.150998	-0.033259	-0.001612	-0.003412	0.016894	0.020775	0.002759	1.000000

Figure 3.6 – The correlation values for our data

We have completed the preprocessing phase of our task, or at least for now. We have removed all irrelevant columns; we also removed the missing values by dropping rows with missing values and, finally, used one-hot encoding to convert our categorical values to numeric values. It is important to note that we are skipping some **exploratory data analysis** (EDA) steps here, such as visualizing the data; although that's an essential step, our core focus for the exam is building models with TensorFlow. In our Colab notebook, you will find some additional EDA steps; although they are not directly relevant to the exams, they will give you a better understanding of your data and help you detect anomalies.

Now, let us move on to the modeling phase.

Model building

To build a model, we will have to sort our data into features (X) and the target (y). To do this, we will run this code block:

```
# We split the attributes and labels into X and y variables
X = df.drop("Salary", axis=1)
y = df["Salary"]
```

We will use the drop() function to drop the Salary column from the X variable, and we will make the y variable the Salary column alone, since this is our target.

With our features and target variable well defined, we can proceed to split our data into training and test sets. This step is important, as it enables our model to learn patterns from our data to effectively predict employees' salaries. To achieve this, we train our model using the training set and then evaluate the model's efficacy on the hold-out test set. We discussed this at a high level in *Chapter 1, Introduction to Machine Learning,* when we talked about the ML life cycle. It is a very important process, as we will use the test set to evaluate our model's generalization capability before we deploy it for real-world use. To split our data into training and testing sets, we will use the sklearn library:

```
X_train, X_test, y_train, y_test = train_test_split(X, y,
    test_size=0.2, random_state=10)
```

Using the train_test_split function from the sklearn library, we split our data into training and testing datasets, with a test size of 0.2. We set the random_state =10 to ensure reproducibility so that every time you use the same random_state value, you'll get the same split, even if you run the code multiple times. For instance, in our code, we set random_state to 10, which means every time we run the code, we will get the same split. If we change this value from 10 to, say, 50, we will get a different shuffled split for our training and test set. Setting the random_state argument when splitting our data into training and test sets is very useful, as it allows us to effectively compare different models, since our training set and test sets are the same across all the models we experiment with.

When modeling our data in machine learning, we usually use 80 percent of the data to train the model and 20 percent of the data to test the model's generalization capability. That's why we set test_size to 0.2 for our dataset. Now that we have everything in place, we will start the modeling process in earnest. When it comes to building models with TensorFlow, there are three essential steps, as illustrated in *Figure 3.7* – building the model, compiling the model, and fitting it to our data.

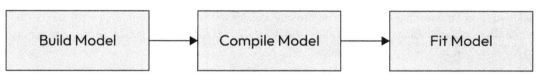

Figure 3.7 – The three-step modeling process

Let us see how we can use this three-step approach to build our salary prediction model. We will begin by building our model:

```
#create a model using the Keras API
Model_1 = Sequential([Dense(units=1, activation='linear',
    input_shape=[len(X_train.columns)])])
```

In *Figure 3.8*, we can see the first line of code for our model. Here, we generated a single layer using the `Sequential` class as an array. The `Sequential` class is used for layer definition. The `Dense` function is used to generate a layer of fully connected neuron. In this case, we have just one unit. For our activation function here, we employ a linear activation function. Activation functions are used to determine the output of a neuron based on a given input or set of inputs. Here, the linear activation function simply outputs whatever the input is – that is, a direct relationship between the input and the output. Next, we pass in the input shape of our data, which in this case is 8, representing the features in data (columns) in `X_train`.

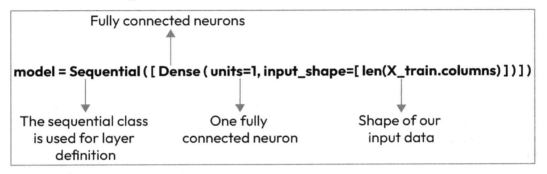

Figure 3.8 – Building a model in TensorFlow

In the first step of our three-step process, we designed the model structure. Now, we will move on to the model compilation step. This step is equally important as it determines how the model will learn. Here, we specify parameters such as the loss function, the optimizer, and the metrics we want to use to evaluate our model.

The optimizer determines how our model will update its internal parameters, based on the information it gathers from the loss function and the data. The job of the loss function is to measure how well our model does on our training data. We then use our metrics to monitor the model's performance on the training step and test steps. Here, we use **stochastic gradient descent** (**SGD**) as our optimizer and MAE for our loss and evaluation metric:

```
#compile the model
Model_1.compile(loss=tf.keras.losses.mae,
    optimizer=tf.keras.optimizers.SGD(), metrics = ['mae'])
```

Now, all we have to do is feed our model with training data and the corresponding labels, with which our model can learn to intelligently predict the target numerical values, which in our case is the expected salary. Every time the model makes a prediction, the loss function compares the difference between the model's prediction and the ground truth. This information is passed to the optimizer, which uses the information to make an improved prediction until the model can fashion the right mathematical equation to accurately predict our employee's salary.

Now, let's fit our training model:

```
#Fit the model
model_1.fit(X_train, y_train, epochs =50)
```

We use model_1.fit to fit our training data and labels and set the number of tries (epochs) to 50. In just a few lines of code, we have generated a mini-brain that we can train over time to make sensible predictions. Let's run the code and see what the output looks like:

```
Epoch 46/50
6/6 [==============================] - 0s 9ms/step - loss: 97378.0391
 - mae: 97378.0391
Epoch 47/50
6/6 [==============================] - 0s 4ms/step - loss: 97377.2500
 - mae: 97377.2500
Epoch 48/50
6/6 [==============================] - 0s 4ms/step - loss: 97376.4609
 - mae: 97376.4609
Epoch 49/50
6/6 [==============================] - 0s 3ms/step - loss: 97375.6484
 - mae: 97375.6484
Epoch 50/50
6/6 [==============================] - 0s 3ms/step - loss: 97374.8516
 - mae: 97374.8516
```

We have displayed the last five tries (epochs 46–50). The error drops gradually; however, we end up with a very large error after 50 epochs. Perhaps we can train our model for more epochs, as we did in *Chapter 2, Introduction to TensorFlow*. Why not?

```
#create a model using the Keras API
model_2 = Sequential([Dense(units=1, activation='linear',
    input_shape=[len(X_train.columns)])])
#compile the model
model_2.compile(loss=tf.keras.losses.mae,
    optimizer=tf.keras.optimizers.SGD(), metrics = ['mae'])
#Fit the model
history=model_2.fit(X_train, y_train, epochs =500)
```

Now, we simply change the number of epochs to 500 using our single-layer model. The activation function, the loss, and optimizers are the same as our initial model:

```
Epoch 496/500
6/6 [==============================] - 0s 3ms/step - loss: 97014.8516
- mae: 97014.8516
Epoch 497/500
6/6 [==============================] - 0s 2ms/step - loss: 97014.0391
- mae: 97014.0391
Epoch 498/500
6/6 [==============================] - 0s 3ms/step - loss: 97013.2500
- mae: 97013.2500
Epoch 499/500
6/6 [==============================] - 0s 3ms/step - loss: 97012.4453
- mae: 97012.4453
Epoch 500/500
6/6 [==============================] - 0s 3ms/step - loss: 97011.6484
- mae: 97011.6484
```

From the last five lines of our output, we can see that the loss is still quite high after 500 epochs. You may wish to experiment with the model for longer epochs to see how it will fare. It is also a good idea to visualize your model's loss curve to see how it performs. A lower loss indicates a better-performing model. With this in mind, let us explore the loss curve for model_2:

```
def visualize_model(history, ymin=None, ymax=None):
    # Lets visualize our model
    print(history.history.keys())
    # Lets plot the loss
    plt.plot(history.history['loss'])
    plt.title('Model loss')
    plt.ylabel('Loss')
    plt.xlabel('Number of epochs')
    plt.ylim([ymin,ymax]) # To zoom in on the y-axis
    plt.legend(['loss plot'], loc='upper right')
    plt.show()
```

We will generate a utility plotting function, visualize_model, which we will use in our experiments to plot the model's loss over time as it trains. In this code, we generate a figure to plot the loss values stored in the history object. The history object is the output of the fit function in our three-step modeling process, and it holds the loss and metrics values at the end of each epoch.

To plot model_2, we simply call the function to visualize the plot and pass in history_2:

```
visualize_model(history_2)
```

When we run the code, we get the plot shown in *Figure 3.9*:

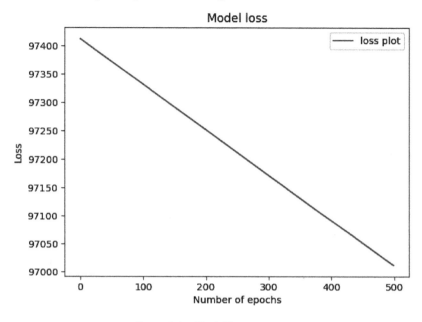

Figure 3.9 – Model losses at 500

From *Figure 3.9*, we can see the loss falling, and the rate at which it falls is too slow, as it takes **500** epochs to fall from around **97400** to **97000**. In your spare time, you can try to train the model for 2,000 or more epochs. It will not be able to generalize well, as the model is too simple to handle the complexity of our data. In machine learning lingo, we say the model is *underfitting*.

> **Note**
>
> There are primarily two ways to build models with TensorFlow – the sequential API and the functional API. The sequential API is a simple way of building models by using a stack of layers, where data flows in a single direction, from the input layer to the output layer. Conversely, the functional API in TensorFlow allows us to build more complex models – this includes models with multiple inputs or outputs and models with shared layers. Here, we use the Sequential API. For more information about building models with the sequential API, check out the documentation: https://www.tensorflow.org/guide/keras/sequential_model.

Hence, let us try to build a more complex model and see whether we can push the loss lower and quicker than our initial model:

```
#Set random set
tf.random.set_seed(10)
```

```
#create a model
model_3 =Sequential([
    Dense(units=64, activation='relu',
    input_shape=[len(X_train.columns)]),
    Dense(units=1)
    ])
#compile the model
model_3.compile(loss="mae", optimizer="SGD",
    metrics = ['mae'])

#Fit the model
history_3 =model_3.fit(X_train, y_train, epochs=500)
```

Here, we have generated a new model. We stack a 64-neuron layer on top of our single-unit layer. We also use a **Rectified Linear Unit (ReLU)** activation function for this layer; its job is to help our model learn more complex patterns in our data and improve computational efficiency. The second layer is our output layer, made up of a single neuron because we have a regression task (predicting a continuous value). Let's run it for 500 epochs and see whether this will make any difference:

```
Epoch 496/500
6/6 [==============================] - 0s 3ms/step - loss: 3651.6785 -
mae: 3651.6785
Epoch 497/500
6/6 [==============================] - 0s 3ms/step - loss: 3647.4753 -
mae: 3647.4753
Epoch 498/500
6/6 [==============================] - 0s 3ms/step - loss: 3722.4863 -
mae: 3722.4863
Epoch 499/500
6/6 [==============================] - 0s 3ms/step - loss: 3570.9023 -
mae: 3570.9023
Epoch 500/500
6/6 [==============================] - 0s 3ms/step - loss: 3686.0293 -
mae: 3686.0293
```

From the last five lines of our output, we can see that there is a significant drop in our loss to around 3686. Let's also plot the loss curve to get a visual understanding as well.

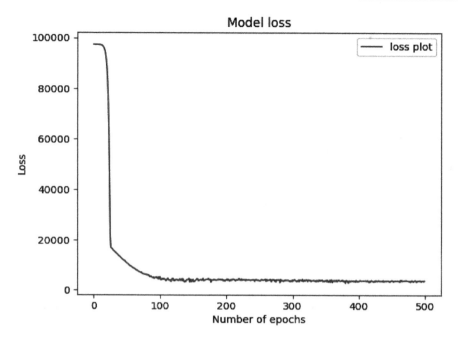

Figure 3.10 – Model losses after 500 epochs

In *Figure 3.10*, we can see that our model's loss has fallen below our lowest recorded loss. This is a massive improvement in comparison to our previous model. However, this is not a desired result, nor does it look like the type of result we would like to present to the HR team. This is because, with this model, if an employee earns $50,000, the model could predict either around $46,300 as the employee's salary, which would make them unhappy, or $53,700 as the employee's salary, in which case the HR team will not be happy. So, we need to figure out how to improve our result.

Let's zoom into the plot to have a better understanding of what is happening with our model:

```
visualize_model(history_3, ymin=0, ymax=10000)
```

When we run the code, it returns the plot shown in *Figure 3.11*.

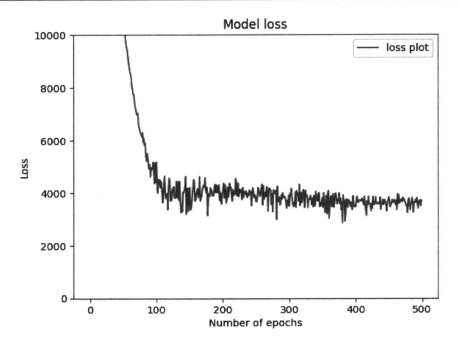

Figure 3.11 – Model losses after 500 epochs when we zoom into the plot

From the plot in *Figure 3.11*, we can see that the loss falls sharply and settles at around the 100th epoch, and nothing significant seems to happen afterward. Hence, training the model for longer just as we did in our previous model may not be the optimal solution. What can we do to improve our model?

Perhaps we can add another layer? Let's do that and see what our results look like. As we initially pointed out, our job requires a lot of experimentation; only then can we learn how to do things better and faster:

```
#Set random set
tf.random.set_seed(10)

#create a model
model_4 =Sequential([
    Dense(units=64, activation='relu',
        input_shape=[len(X_train.columns)]),
    Dense(units=64, activation='relu'),
        Dense(units=1)
    ])
#compile the model
model_4.compile(loss="mae", optimizer="SGD",
    metrics = "mae")
```

```
#fit the model
history_4 =model_4.fit(X_train, y_train, epochs=500)
```

Here, we added another dense layer of 64 neurons. Note that we also use ReLU as the activation function here:

```
Epoch 496/500
6/6 [==============================] - 0s 3ms/step - loss: 97384.4141
 - mae: 97384.4141
Epoch 497/500
6/6 [==============================] - 0s 3ms/step - loss: 97384.3516
 - mae: 97384.3516
Epoch 498/500
6/6 [==============================] - 0s 3ms/step - loss: 97384.3047
 - mae: 97384.3047
Epoch 499/500
6/6 [==============================] - 0s 3ms/step - loss: 97384.2422
 - mae: 97384.2422
Epoch 500/500
6/6 [==============================] - 0s 3ms/step - loss: 97384.1797
 - mae: 97384.1797
```

We display only the last five epochs, and we can see the loss is around 97384, which is worse than the results achieved in model_3. So, how do we know how many layers to use in our modeling process? The answer is by experimenting. We use trial and error, backed by our understanding of what the results look like. We can decide whether we need to add more layers, as we did initially when the model was underfitting. And should the model get so complex that it masters the training data well but does not generalize well on our test (hold-out) data, it is said to be *overfitting* in machine learning lingo.

Now that we have tried smaller and larger models, we cannot yet say we have achieved a suitable result, and the HR manager has checked in on us to find out how we are doing in terms of the prediction modeling task. So far, we did some research, as all ML engineers do, and we discovered a very important step that we can try out. What step? Let's see.

Normalization

Normalization is a technique applied to input features to ensure they are of a consistent scale, usually between 0 and 1. This process helps our model to converge faster and more accurately. It is worth noting that we should apply normalization after completing other data preprocessing steps, such as handling missing values.

It's good practice to know that improving your model output can also rely strongly on your data preparation process. Hence, let us apply this here. We will take a step back from model building and look at our features after we converted all the columns to numerical values:

```
X.describe()
```

We will use the `describe` function to get vital statistics of our data. This information shows us that most of the columns have a minimum value of 0 and a maximum value of 1, but the experience column is of a different scale, as shown in *Figure 3.12*:

	Experience	Qualification_Msc	Qualification_PhD	University_Tier2	University_Tier3	Role_Mid	Role_Senior	Cert_Yes
count	241.000000	241.000000	241.000000	241.000000	241.000000	241.000000	241.000000	241.000000
mean	3.427386	0.294606	0.240664	0.377593	0.248963	0.215768	0.518672	0.518672
std	1.493010	0.456814	0.428376	0.485794	0.433312	0.412210	0.500691	0.500691
min	1.000000	0.000000	0.000000	0.000000	0.000000	0.000000	0.000000	0.000000
25%	2.000000	0.000000	0.000000	0.000000	0.000000	0.000000	0.000000	0.000000
50%	4.000000	0.000000	0.000000	0.000000	0.000000	0.000000	1.000000	1.000000
75%	5.000000	1.000000	0.000000	1.000000	0.000000	0.000000	1.000000	1.000000
max	5.000000	1.000000	1.000000	1.000000	1.000000	1.000000	1.000000	1.000000

Figure 3.12 – A statistical summary of the dataset (before normalization)

Why does this matter, you may ask? When the scale of our data is different, our model will arbitrarily attach more importance to columns with higher values, which could affect the model's ability to predict our target correctly. To resolve this issue, we will use normalization to scale the data between 0 and 1 to bring all our features to the same scale, thereby giving every feature an equal chance when our model begins to learn how they relate to our target (y).

To normalize our data, we will use the following equation to scale it:

$$X_{norm} = \frac{X - X_{min}}{X_{max} - X_{min}}$$

where X is our data, X_{min} is the minimum value of X, and X_{max} is the maximum value of X. In our case study, the minimum value of X in the `Experience` column is 1, and the maximum value of X in the `Experience` column is 7. The good news is that we can easily implement this step using the `MinMaxScaler` function from the `sklearn` library. Let's see how to scale our data next:

```
# create a scaler object
scaler = MinMaxScaler()
# fit and transform the data
X_norm = pd.DataFrame(scaler.fit_transform(X),
    columns=X.columns)

X_norm.describe()
```

Let's use the `describe()` function to view the key statistics again, as shown in *Figure 3.13*.

	Experience	Qualification_Msc	Qualification_PhD	University_Tier2	University_Tier3	Role_Mid	Role_Senior	Cert_Yes
count	241.000000	241.000000	241.000000	241.000000	241.000000	241.000000	241.000000	241.000000
mean	0.606846	0.294606	0.240664	0.377593	0.248963	0.215768	0.518672	0.518672
std	0.373253	0.456814	0.428376	0.485794	0.433312	0.412210	0.500691	0.500691
min	0.000000	0.000000	0.000000	0.000000	0.000000	0.000000	0.000000	0.000000
25%	0.250000	0.000000	0.000000	0.000000	0.000000	0.000000	0.000000	0.000000
50%	0.750000	0.000000	0.000000	0.000000	0.000000	0.000000	1.000000	1.000000
75%	1.000000	1.000000	0.000000	1.000000	0.000000	0.000000	1.000000	1.000000
max	1.000000	1.000000	1.000000	1.000000	1.000000	1.000000	1.000000	1.000000

Figure 3.13 – A statistical summary of the dataset (after normalization)

Now, all our data is of the same scale. So, we have successfully implemented normalization of our data in just a few lines of code.

Now, we split our data into training and testing sets, but this time, we use our normalized X (X_norm) in the code:

```
# Create training and test sets with the normalized data (X_norm)
X_train, X_test, y_train, y_test = train_test_split(X_norm,
    y,  test_size=0.2, random_state=10)
```

Now, we use our best-performing model (model_3) from the initial experiments we have done so far. Let's see how our model performs after normalization:

```
#create a model
model_5 =Sequential([
    Dense(units=64, activation='relu',
        input_shape=[len(X_train.columns)]),
    Dense(units=64, activation ="relu"),
        Dense(units=1)
    ])
#compile the model
model_5.compile(loss="mae",
    optimizer=tf.keras.optimizers.SGD(), metrics = ['mae'])
history_5 =model_5.fit(X_train, y_train, epochs=1000)
```

The output is as follows:

```
Epoch 996/1000
6/6 [==============================] - 0s 4ms/step - loss: 1459.2953 -
mae: 1459.2953
Epoch 997/1000
6/6 [==============================] - 0s 4ms/step - loss: 1437.8248 -
mae: 1437.8248
Epoch 998/1000
```

```
6/6 [==============================] - 0s 3ms/step - loss: 1469.3732 -
mae: 1469.3732
Epoch 999/1000
6/6 [==============================] - 0s 4ms/step - loss: 1433.6071 -
mae: 1433.6071
Epoch 1000/1000
6/6 [==============================] - 0s 3ms/step - loss: 1432.2891 -
mae: 1432.2891
```

From the results, we can see that MAE has reduced by more than half in comparison to the results we got without applying normalization.

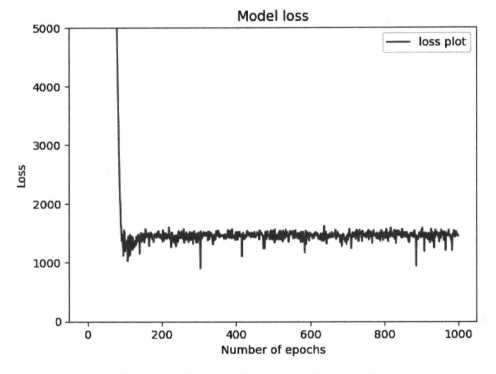

Figure 3.14 – The zoomed-in loss curve for model_5

Also, if you look at the loss plot for model_5 in *Figure 3.14*, you can see the loss fails to drop significantly after around the 100th epoch. Instead of guessing how many epochs are ideal to train the model, how about we set a rule to stop training when the model fails to improve its performance? Also, we can see that model_5 doesn't give us the result we want; perhaps now is a good time to try out a bigger model, in which we train it for longer and set a rule to stop training once it fails to improve its performance on the training data:

```
#create a model

model_6 =Sequential([
    Dense(units=64, activation='relu',
        input_shape=[len(X_train.columns)]),
        Dense(units=64, activation ="relu"), Dense(units=1)
    ])
#compile the model
model_6.compile(loss="mae",
    optimizer=tf.keras.optimizers.SGD(), metrics = ['mae'])
#fit the model
early_stop=keras.callbacks.EarlyStopping(monitor='loss',
    patience=10)
history_6 =model_6.fit(
    X_train, y_train, epochs=1000, callbacks=[early_stop])
```

Here, we use a three-layer model; the first two layers are made up of 64 neurons and the output layer has a single neuron. To set the rule to stop training, we use *early stopping*; this additional parameter is applied when we fit our model on the data to stop training when the model loss fails to improve after 10 epochs. This is achieved by specifying the metric to monitor loss and setting `patience` to `10`. Early stopping is also a great technique to prevent overfitting, as it stops training when the model fails to improve; we will discuss this further in *Chapter 6, Improving the Model*. Let's look at the result now:

```
Epoch 25/1000
6/6 [==============================] - 0s 3ms/step - loss: 84910.6953
- mae: 84910.6953
Epoch 26/1000
6/6 [==============================] - 0s 3ms/step - loss: 81037.8516
- mae: 81037.8516
Epoch 27/1000
6/6 [==============================] - 0s 3ms/step - loss: 72761.0078
- mae: 72761.0078
Epoch 28/1000
6/6 [==============================] - 0s 3ms/step - loss: 81160.6562
- mae: 81160.6562
Epoch 29/1000
6/6 [==============================] - 0s 3ms/step - loss: 70687.3125
- mae: 70687.3125
```

Although we set our training for 1000 epochs, our EarlyStopping callback halted the training process on the 29th epoch because it observed no meaningful drop in the loss. Although the result here isn't great, we have used EarlyStopping to save a considerable amount of computational resources and time. Perhaps now is a good time to try out a different optimizer. For this next experiment, let's use the Adam optimizer. Adam is another popular optimizer that is used in deep learning, due to its ability to adaptively control the learning rate for each parameter in a model, which accelerates the model's convergence:

```
#create a model
model_7 =Sequential([
    Dense(units=64, activation='relu',
        input_shape=[len(X_train.columns)]),
    Dense(units=64, activation ="relu"),
    Dense(units=1)


    ])
#compile the model
model_7.compile(loss="mae", optimizer="Adam",
    metrics ="mae")
#fit the model
early_stop=keras.callbacks.EarlyStopping(monitor='loss',
    patience=10)
history_7 =model_7.fit(
    X_train, y_train, epochs=1000, callbacks=[early_stop])
```

Note we only changed the optimizer to Adam in our compile step. Let's see the result of this change in the optimizer:

```
Epoch 897/1000
6/6 [==============================] - 0s 4ms/step - loss: 30.4748 -
mae: 30.4748
Epoch 898/1000
6/6 [==============================] - 0s 4ms/step - loss: 19.4643 -
mae: 19.4643
Epoch 899/1000
6/6 [==============================] - 0s 3ms/step - loss: 17.0965 -
mae: 17.0965
Epoch 900/1000
6/6 [==============================] - 0s 3ms/step - loss: 18.5009 -
mae: 18.5009
Epoch 901/1000
6/6 [==============================] - 0s 3ms/step - loss: 15.5516 -
mae: 15.5516
```

By just changing the optimizer, we have recorded an incredible drop in loss. Also, note that we did not need the entire 1000 epochs, as training ended on 901 epochs. Let us add another layer; perhaps we will see an improved performance:

```
#create a model
model_8 =Sequential([
    Dense(units=64, activation='relu',
        input_shape=[len(X_train.columns)]),
    Dense(units=64, activation ="relu"),
    Dense(units=64, activation ="relu"),
    Dense(units=1)


    ])
#compile the model
model_8.compile(loss="mae", optimizer="Adam",
    metrics ="mae")
#fit the model
early_stop=keras.callbacks.EarlyStopping(monitor='loss',
    patience=10)
history_8 =model_8.fit(
    X_train, y_train, epochs=1000, callbacks=[early_stop])
```

Here, we added an extra layer with 64 neurons, with ReLU as the activation function. Everything else is the same:

```
Epoch 266/1000
6/6 [==============================] - 0s 4ms/step - loss: 73.3237 -
mae: 73.3237
Epoch 267/1000
6/6 [==============================] - 0s 4ms/step - loss: 113.9100 -
mae: 113.9100
Epoch 268/1000
6/6 [==============================] - 0s 4ms/step - loss: 257.4851 -
mae: 257.4851
Epoch 269/1000
6/6 [==============================] - 0s 4ms/step - loss: 149.9819 -
mae: 149.9819
Epoch 270/1000
6/6 [==============================] - 0s 4ms/step - loss: 179.7796 -
mae: 179.7796
```

Training stops at 270 epochs; although our model is more complex, it doesn't perform better than model_7 on training. We have tried out different ideas while experimenting; now, let us try out all eight models on the test set and evaluate them.

Model evaluation

To evaluate our models, we will write a function to apply the `evaluate` metrics to all eight models:

```
def eval_testing(model):
    return model.evaluate(X_test, y_test)
models = [model_1, model_2, model_3, model_4, model_5,
    model_6, model_7, model_8]
for x in models:
    eval_testing(x)
```

We will generate an `eval_testing(model)` function that takes a model as an argument and uses the `evaluate` method to evaluate the performance of the model on our test dataset. Looping through the list of models, the code returns the loss and MAE values for all eight models for our test data:

```
2/2 [==============================] - 0s 8ms/step - loss: 100682.4609
- mae: 100682.4609
2/2 [==============================] - 0s 8ms/step - loss: 100567.9453
- mae: 100567.9453
2/2 [==============================] - 0s 10ms/step - loss: 17986.0801
- mae: 17986.0801
2/2 [==============================] - 0s 9ms/step - loss: 100664.0781
- mae: 100664.0781
2/2 [==============================] - 0s 6ms/step - loss: 1971.4187 -
mae: 1971.4187
2/2 [==============================] - 0s 11ms/step - loss: 5831.1250
- mae: 5831.1250
2/2 [==============================] - 0s 7ms/step - loss: 5.0099 -
mae: 5.0099
2/2 [==============================] - 0s 26ms/step - loss: 70.2970 -
mae: 70.2970
```

After we evaluate the models, we can that see `model_7` has the lowest loss. Let's see how it does on our test set by using it to make predictions.

Making predictions

Now that we are done with experimenting and have evaluated the models, let's use `model_7` to predict our test set salaries and see how they compare with the ground truth. To do this, we will use the `predict()` function:

```
#Let's make predictions on our test data
y_preds=model_7.predict(X_test).flatten()
y_preds
```

After we run this code block, we get the output in an array, as shown here:

```
2/2 [==============================] - 0s 9ms/step
array([ 64498.64 , 131504.89 , 116491.73 ,  72500.13 , 102983.836,
        60504.645,  84503.36 , 119501.664, 112497.734,  63501.168,
        77994.87 ,  84497.16 , 112497.734,  90980.625,  87499.88 ,
       100502.234, 135498.88 , 112491.53 , 119501.664, 131504.89 ,
       108990.31 , 117506.63 ,  80503.16 , 123495.66 , 112497.734,
       117506.63 , 111994.03 ,  78985.125, 135498.88 , 129502.125,
       117506.64 , 119501.664, 100502.234, 113506.43 , 101987.38 ,
       113506.43 ,  93990.555,  65496.2  ,  61494.906, 107506.17 ,
       105993.77 , 106502.5  ,  72493.94 , 135498.88 ,  67501.37 ,
       107506.17 , 117506.63 ,  70505.1  ,  57500.906], dtype=float32)
```

For clarity, let's build a DataFrame with the model's prediction and ground truth. This should be fun and somewhat magical when you see how good our model has become:

```
#Let's make a DataFrame to compare our prediction with the ground
truth
df_predictions = pd.DataFrame({'Ground_Truth': y_test,
    'Model_prediction': y_preds}, columns=['Ground_Truth',
    'Model_prediction']) df_predictions[
    'Model_prediction']= df_predictions[
    'Model_prediction'].astype(int)
```

Here, we generate two columns and convert the model's prediction from `float` to `int`, just to keep it in scope with the ground truth. Ready for the result?

We will use the `head` function to print out the first 10 values of the test set:

```
#Let's look at the top 10 data points in the test set
df_predictions.sample(10)
```

We then see our results, as shown in *Figure 3.15*:

	Ground_Truth	Model_prediction	diff
31	123500	123495	5
116	80500	80503	-3
104	94000	93990	10
121	63500	63501	-1
81	79000	78985	15
233	112500	112497	3
101	135500	135498	2
135	131500	131504	-4
9	112500	112497	3
246	107500	107506	-6

Figure 3.15 – A DataFrame showing the actual values, predictions made by the model, and the resulting residuals

Our model has achieved something impressive; it is really close to the initial salaries in our test data. Now, you can show the HR manager your amazing result. We must save the model so that we can load it and make predictions anytime we want. Let's learn how to do this next.

Saving and loading models

The beauty of TensorFlow is the ease with which we can do complex stuff. To save a model, we just need one line of code:

```
#Saving the model in one line of code
Model7.save('salarypredictor.h5')
#Alternate method is
#model7.save('salarypredictor')
```

You can save it as `your_model.h5` or `your_model`; either way works. TensorFlow recommends the `SavedModel` approach because it is language-agnostic, which makes it easy to deploy on various platforms. In this format, we can save the model and its individual components, such as the weights and variables. Conversely, the HDF5 format saves the complete model structure, its weights, and the training configurations as a single file. This approach gives us greater flexibility to share and distribute models; however, for deployment purposes, it's not the preferred method. When we run the code, we can see the saved model on the left-hand panel in our Colab notebook, as shown in *Figure 3.16*.

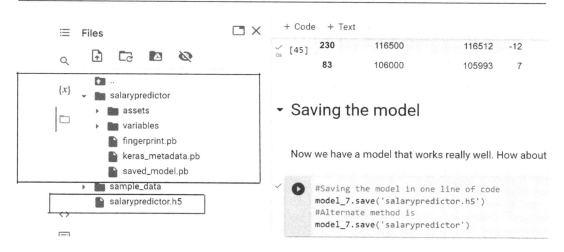

Figure 3.16 – A snapshot of our saved model

Now that we have saved the model, it is wise to test it out by reloading it and testing it. Let's do that. Also, it's just one line of code to load the model:

```
#loading the model
saved_model =tf.keras.models.load_model("/content/salarypredictor.h5")
```

Let's try out our `saved_model` and see whether it will work as well as `model_7`. We will generate `y_pred` again and generate a DataFrame, using `y_test` and `y_pred` as we did earlier checking first the 10 random samples from our test data:

	Ground_Truth	Model_prediction
246	107500	107506
228	119500	119501
181	60500	60504
216	72500	72500
18	102000	101987
217	91000	90980
125	106500	106502
15	112500	112491
90	87500	87499
134	113500	113506

Figure 3.17 – A DataFrame showing the actual values and predictions made by the saved model

From the results in *Figure 3.17*, we can see that our saved model performs at a high level. Now, you can deliver your result to the HR manager, and they should be excited about your results. Let's imagine that the HR manager wants you to use your model to predict the salary of the new hires. Let's do that next:

```
#Putting everything into a function for our big task
def salary_predictor(df):
    df_hires= df.drop(columns=['Name', 'Phone_Number',
        'Date_Of_Birth' ])
    df_hires = pd.get_dummies(df_hires, drop_first=True)
    X_norm = pd.DataFrame(scaler.fit_transform(df_hires),
        columns=df.columns)
    y_preds=saved_model.predict(X_norm).flatten()
    df_predictions = pd.DataFrame({ 'Model_prediction':
        y_preds}, columns=[ 'Model_prediction'])
    df_predictions['Model_prediction']= df_predictions[
        'Model_prediction'].astype(int)
    df['Salary']=df_predictions['Model_prediction']
    return df
```

We generate a function using our saved model. We simply wrap all the steps we've covered so far into the function, and we return a DataFrame. Now, let's read in the data of our new hires:

```
#Load the data
df_new=pd.read_csv('https://raw.githubusercontent.com/oluwole-packt/
datasets/main/new_hires.csv')
df_new
```

When we run the code block, we can see their data in a DataFrame, as shown in *Figure 3.18*.

	Name	Phone_Number	Experience	Qualification	University	Role	Cert	Date_Of_Birth
0	Alvaro Johnson	320-636-8883	7	Bsc	Tier1	Senior	No	12/03/1978
1	Austin Powers	903-121-1691	2	Msc	Tier1	Mid	Yes	13/03/1992
2	Joshua Phil	673-972-2453	3	Bsc	Tier3	Mid	Yes	19/02/1988
3	Mirinda Collins	310-364-6925	5	Msc	Tier2	Senior	No	20/03/1989
4	Mustapha Green	401-249-3912	3	PhD	Tier1	Junior	Yes	21/03/1979
5	Nick Freeman	875-546-2104	6	Bsc	Tier3	Junior	Yes	22/03/1982
6	Pamela Allison	408-955-5085	2	PhD	Tier2	Junior	No	23/03/1968

Figure 3.18 – A DataFrame showing the new hires

Now, we pass the data into the function we generated to get the predicted salaries for our new hires:

```
#Lets see how much
salary_predictor(df_new)
```

We pass `df_new` into the salary prediction function, and we get a new DataFrame, as shown in *Figure 3.19*:

	Name	Phone_Number	Experience	Qualification	University	Role	Cert	Date_Of_Birth	Salary
0	Alvaro Johnson	320-636-8883	7	Bsc	Tier1	Senior	No	12/03/1978	100481
1	Austin Powers	903-121-1691	2	Msc	Tier1	Mid	Yes	13/03/1992	85976
2	Joshua Phil	673-972-2453	3	Bsc	Tier3	Mid	Yes	19/02/1988	87170
3	Mirinda Collins	310-364-6925	5	Msc	Tier2	Senior	No	20/03/1989	110084
4	Mustapha Green	401-249-3912	3	PhD	Tier1	Junior	Yes	21/03/1979	83681
5	Nick Freeman	875-546-2104	6	Bsc	Tier3	Junior	Yes	22/03/1982	79265
6	Pamela Allison	408-955-5085	2	PhD	Tier2	Junior	No	23/03/1968	77499

Figure 3.19 – A DataFrame showing the new hires with their predicted salaries

Finally, we have achieved our goal. HR is happy, the new hires are happy, and everyone in the company thinks you are a magician. Perhaps a pay raise could be on the table, but while you bask in the euphoria around your first success, your manager returns with another task. This time, it is a classification task, which we will look at this in the next chapter. For now, good job!

Summary

In this chapter, we took a deeper dive into supervised learning, with a focus on regression modeling. Here, we discussed the difference between simple and multiple linear regression and looked at some important evaluation metrics for regression modeling. Then, we rolled up our sleeves on our case study, helping our company build a working regression model to predict the salaries of new employees. We carried out some data preprocessing steps and saw the importance of normalization in our modeling process.

At the end of the case study, we successfully built a salary prediction model, evaluated the model on our test set, and mastered how to save and load models for use at a later stage. Now, you can confidently build a regression model with TensorFlow.

In the next chapter, we'll take a look at classification modeling.

Questions

Let's test what we learned in this chapter.

1. What is linear regression?
2. What is the difference between simple and multiple linear regression?
3. What evaluation metric penalizes large errors in regression modeling?
4. Use the salary dataset to forecast salaries.

Further reading

To learn more, you can check out the following resources:

- Amr, T., 2020. *Hands-On Machine Learning with scikit-learn and Scientific Python Toolkits.* [S.l.]: Packt Publishing.

- Raschka, S. and Mirjalili, V., 2019. *Python Machine Learning.* 3rd ed. Packt Publishing.

- *TensorFlow documentation*: https://www.TensorFlow.org/guide.

4
Classification with TensorFlow

In the last chapter, we covered linear regression with TensorFlow, where we looked at both simple and multiple linear regression; we also explored various metrics for evaluating regression models. We concluded the chapter with a real-world use case, where we built a salary prediction model, and we used this to predict the salaries of new employees based on a set of features. In this chapter, we will continue with modeling in TensorFlow – this time, by exploring classification problems with TensorFlow.

We will start by looking at the concept of classification modeling, after which we will examine the various evaluation metrics for classification modeling and how we can apply them to various use cases. We will look at binary, multi-class, and multi-label classification modeling. Finally, we will walk through a case study, putting all we have learned into practice by building a binary classification model to predict whether a student will drop out of university or not.

By the end of this chapter, you should clearly understand what classification modeling in machine learning is and also be able to differentiate between binary, multi-class, and multi-label classification problems. You will be familiar with how to build, compile, train, predict, and evaluate classification models.

In this chapter, we'll cover the following topics:

- Classification with TensorFlow
- A student dropout prediction

Technical requirements

In this chapter, we will use Google Colab to run the coding exercise, and you will need to install Python >= 3.8.0, along with the following packages, which can be installed using the `pip install` command:

- `tensorflow >= 2.7.0`
- `tensorflow-datasets == 4.4.0`
- `Pillow == 8.4.0`
- `pandas == 1.3.4`

- `numpy == 1.21.4`
- `scipy == 1.7.3`

The code bundle for this book is available at the following GitHub link: `https://github.com/PacktPublishing/TensorFlow-Developer-Certificate`. Solutions to all the exercises can also be found at this link.

Classification with TensorFlow

In *Chapter 1, Introduction to Machine Learning*, we talked about supervised learning and briefly talked about classification modeling. Classification modeling involves predicting classes in our target variable. When the classes we try to predict are binary (for example, trying to predict whether a pet is either a dog or a cat, whether an email is spam or not, or whether a patient has cancer or not), this type of classification scenario is referred to as **binary classification**.

Then again, we may be faced with a problem where we want to build an ML model to predict the different breeds of dogs. In this case, we have more than two classes, so this type of classification is called **multi-class classification**. Just like binary classification problems, in multi-class classification, our target variable can only belong to one class out of multiple classes – our model will select either a bulldog, a German shepherd, or a pit bull. Here, the classes are *mutually exclusive*.

Imagine that you are building a movie classifier, and you want to classifier a blockbuster movie such as *Avengers: Endgame*. This movie belongs to the action, adventure, superhero, epic, fantasy, and science fiction genres. From the movie's label, we can see that our target variable belongs to more than one genre; hence, this type of classification is called **multi-label classification**, where the output class has more than one target label.

Unlike in multi-class classification, where each example can only belong to one class, in multi-label classification, each example can belong to multiple labels.

Unlike binary and multi-class, where each example can only belong to one class, in multi-label classification, each example can belong to multiple classes, just like the *Avengers* movie. Now that we have looked at the three main types of classification problems, the next question is, how do we evaluate classification models? What are the key metrics we need to look out for? Let us look at this now and understand what they mean and how to best apply them to various classification problems.

Evaluating classification models

Unlike regression problems, where we have numeric values in our target variable, in classification modeling, we have established that our output is classes. Hence, we cannot use the same metrics we used to evaluate our regression models in *Chapter 3, Linear Regression with TensorFlow*, since our output is not continuous numerical values but classes. For a classification problem, let's say we build a spam filtering system to classify a client's emails. The client has 250 emails that are not spam and another 250 emails that are spam. Using our spam filtering model, we are able to correctly flag 230 spam messages and also correctly identify 220 non-spam messages as not spam.

When our spam filter correctly identifies a spam message as spam (which is what we want), we call this a **true positive**, and when the model misclassifies a spam message as not spam, this is called a **false negative**. In a case where the model correctly identifies a non-spam email as not spam, this is called a **true negative**; however, we occasionally find important emails in our spam folder, and these messages were wrongly filtered as spam when they were not. This scenario is called a **false positive**. We can now use these details to evaluate the performance of our spam-filtering model.

Let us list the important details we now know:

- **Total spam messages**: 250 samples
- **Correctly predicted spam messages (true positives)**: 230 samples
- **Wrongly predicted spam messages (false negatives or a type 2 error)**: 20 samples
- **Total non-spam messages**: 250 samples
- **Correctly predicted non-spam messages (true negatives)**: 220 samples
- **Wrongly predicted spam messages (false positive or a type 1 error)**: 30 samples

Now that we have gathered the key details, let us now use them to learn how to evaluate classification models. To do this, we will have to talk about the confusion matrix next.

Confusion matrix

The **confusion matrix** is an error matrix that displays the performance of a classification model in a tabular form, containing both the true values and the predicted values, as illustrated in *Figure 4.1*.

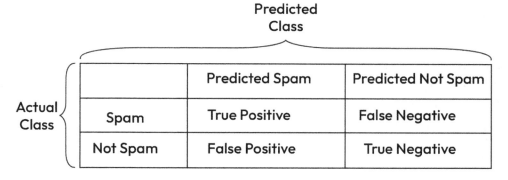

		Predicted Class	
		Predicted Spam	Predicted Not Spam
Actual Class	Spam	True Positive	False Negative
	Not Spam	False Positive	True Negative

Figure 4.1 – The confusion matrix

Using the confusion matrix, we can calculate various classification evaluation metrics such as accuracy, precision, recall, and F1 score. In the confusion matrix, we can see the predicted class at the top, showing emails predicted as spam in the first column and those predicted as not spam in the second column, while the rows show us the true values. Here, we can see in the first row the true spam class and the true not-spam class. When we put it all together, we can see the true values and the wrong prediction in a tabular fashion, which gives us a quick view of the model and its performance across both classes. Let's use these details to compute key performance metrics for our model.

Accuracy is quite intuitive, as it is the sum of the correctly predicted labels over the total available data. We can represent this with the following equation:

$$Accuracy = \frac{TP + TN}{(TP + FP + TN + FN)}$$

Let's add our values and see what our accuracy will look like:

$$Accuracy = \frac{230 + 220}{(230 + 30 + 220 + 20)} = 0.90$$

We get an accuracy of 90%. This is potentially exciting, but let us be more realistic with our data. When it comes to spam emails, we will likely have more legitimate emails than spam emails coming into our mailbox.

Let's imagine we have another client, B, with 500 emails, made up of the following details:

- **Total spam messages**: 40 samples
- **Correctly predicted spam messages (true positives)**: 20 samples
- **Wrongly predicted spam messages (false negatives or a type 2 error)**: 20 samples
- **Total non-spam messages**: 460 samples
- **Correctly predicted non-spam messages (true negatives)**: 430 samples
- **Wrongly predicted spam messages (false positive or a type 1 error)**: 30 samples

If we compute the accuracy for client B, we have:

$$Accuracy = \frac{20 + 430}{20 + 30 + 430 + 20} = 0.90$$

Again, we arrive at an accuracy of 90 percent, yet our model could only predict 50 percent of the spam emails as spam. This shows us that accuracy may not always be the best measure, especially when we deal with a use case made up of imbalanced data such as email classification, fraud detection, or disease detection.

To get a better sense of how our model is doing, we will now turn our attention to precision and recall. Referring back to *Figure 4.2*, the ratio of the true positive to the positive class in the ground truth is called *sensitivity* or *recall*, or the *true positive rate* in ML lingo, and it is represented by the following equation:

$$Recall = \frac{TP}{(TP + FN)}$$

Whereas *precision* is the ratio of the true positive to the positive class predicted by the model, and we also represent it as an equation:

$$Precision = \frac{TP}{(TP + FP)}$$

Using client B, let us calculate our model's performance using precision and recall:

$$Precision\ for\ case\ study\ 2 = \frac{20}{(20 + 30)} = 0.4$$

$$Recall\ for\ case\ study\ 2 = \frac{20}{(20 + 20)} = 0.5$$

Now, we can see how badly our model is, although it has a high accuracy on the entirety of the data. Another important metric that we will come across for classification tasks is the F1 score. The *F1 score* combines recall and precision, and we arrive at it by computing the harmonic mean of precision and recall:

$$F1\ Score = 2 * \frac{precision * recall}{(precision + recall)}$$

Let us calculate the F1 score for the second case study:

$$F1\ Score = 2 * \frac{0.4 * 0.5}{(0.4 + 0.5)} = 0.44$$

From our evaluation of Client B's emails using our spam-filtering model, we now know we need to build a more effective model, one with much better precision and recall for our target class. However, achieving high precision and recall may not always be possible. In such a scenario, we are left with a trade-off, which is known as the *precision/recall trade-off*. In the case of detecting spam emails, we know clients are unlikely to switch to a different service provider should a few spam messages find their way into their inbox; however, they will be upset if they fail to find important messages in their inbox. In this instance, we will aim to achieve a higher recall. Conversely, let's say we build an early cancer detection system, where our focus will be on achieving high precision to minimize false positives. It is important to note that precision and recall are not mutually exclusive, and we can achieve both high precision and recall with a well-tuned model in many instances.

We have now covered some important classification metrics. Now, let us look at a case study (a student dropout prediction) where we will build and evaluate our classification models using different modules from TensorFlow and scikit-learn. Let's jump in.

A student dropout prediction

In *Chapter 3, Linear Regression with TensorFlow*, you began your journey using TensorFlow to build a salary prediction model. Your boss was impressed, and now that you are fully settled in the data team, your manager wants you to work with a new client. Your job is to help them build a model that will predict whether a student will drop out of university or not, as this will help them support such students, thus preventing them from dropping out of school. Your manager has given you authorization and the task is now yours. For this task, historical data was made available to you by your client. Just like in *Chapter 3, Linear Regression with TensorFlow*, you had a rewarding chat with the client, and you identified the task as a binary classification problem. Let's open the notebook labeled `Classification with TensorFlow` from the GitHub repository and get started.

Loading the data

Let's start by loading the historical data that we received from our client:

1. We will start by importing the TensorFlow libraries that we will use to execute our task:

```
# import tensorflow
import tensorflow as tf
from tensorflow import keras
from tensorflow.keras import Sequential
from tensorflow.keras.layers import Dense
print(tf.__version__)
```

After running the code, we can see the version of TensorFlow we will use. In my case, it's 2.8.0 at the time of writing. You will most likely have a newer version, but it should work just as well:

```
2.8.0
```

2. Then, we will import some additional libraries that will help us simplify our workflow.

```
#import additional libraries
import numpy as np
import pandas as pd
# For visualizations
import matplotlib.pyplot as plt
import seaborn as sns
#for splitting the data into training and test set
from sklearn.model_selection import train_test_split
# For Normalization
from sklearn.preprocessing import MinMaxScaler
# Confusion matrix
from sklearn.metrics import confusion_matrix, classification_
report
```

We previously discussed most of the libraries that we will use here, except for the last line of the code block, which we will use to import the confusion matrix and classification report from the scikit-learn library. We will use these functions to evaluate our model's performance. If you are unclear about the other libraries, refer to *Chapter 3, Linear Regression with TensorFlow*, before proceeding with this case study.

3. Now that we have loaded all the necessary libraries, let us make a DataFrame for easy processing:

```
#Loading data from the course GitHub repository
df=pd.read_csv('https://raw.githubusercontent.com/
PacktPublishing/TensorFlow-Developer-Certificate/main/
Chapter%204/Students-Dropout-Prediction.csv', index_col=0)
df.head()
```

When we run the code, if everything works as expected, we should get the first five rows of our dataset.

	Student ID	Student Name	Library	Resources	Finance	Scholarships	Study Time	Study Group	GPA	Test	Assignment	Graduated
0	1324	David Abbott	Average	Online	Paid	No	Average	Tier 3	2.07	5.1	9.1	Drop out
1	1325	John Ward	Poor	Online	Unpaid	No	Excellent	Tier 1	1.96	10.9	7.8	Drop out
2	1326	Sarah James	Good	Online	Paid	No	Poor	Tier 2	3.77	9.3	6.6	Graduated
3	1327	John Zhang	Good	Offline	Paid	No	Average	Tier 1	1.82	14.6	6.2	Drop out
4	1328	Holly Leonard	Poor	Hybrid	Paid	No	Poor	Tier 1	3.22	14.6	10.9	Graduated

Figure 4.2 – A DataFrame showing the first five rows of our dataset

From the output, we can see that our data is made up of numerical and categorical columns. Each of the rows represents a student. Upon inspection, we can see that we have 12 columns, namely, Student ID, Student Name, Library, Resources, Finance, Scholarships, Study Time, Study Group, GPA, Test, Assignment, and Graduated. To efficiently model our data, we need to do some data preparation, so let's start with some exploratory data analysis and see what we can find.

Exploratory data analysis

Perform the following steps to explore and analyze the data:

1. We will begin the exploratory data analysis process using the df.info() function to check for NULL values as well as the data types in our dataset, as shown in *Figure 4.3*.

```
<class 'pandas.core.frame.DataFrame'>
Int64Index: 25000 entries, 0 to 24999
Data columns (total 12 columns):
 #   Column        Non-Null Count   Dtype
---  ------        --------------   -----
 0   Student ID    25000 non-null   int64
 1   Student Name  25000 non-null   object
 2   Library       25000 non-null   object
 3   Resources     25000 non-null   object
 4   Finance       25000 non-null   object
 5   Scholarships  25000 non-null   object
 6   Study Time    25000 non-null   object
 7   Study Group   25000 non-null   object
 8   GPA           25000 non-null   float64
 9   Test          25000 non-null   float64
 10  Assignment    25000 non-null   float64
 11  Graduated     25000 non-null   object
dtypes: float64(3), int64(1), object(8)
memory usage: 2.5+ MB
```

Figure 4.3 – Information about our dataset

The good news is that we have no missing values in our dataset, and yes, we will work with a much larger dataset than in our regression task. Here, we have 25,000 data points representing students' data collected from the university.

2. The next step is to drop the irrelevant columns. By inspecting the available columns, we drop the student ID and student name columns, as these columns should have no impact on whether a student will graduate or not. Let's do that here, using the drop function from pandas:

```
df = df.drop(['Student ID', 'Student Name'], axis=1)
```

3. Next, let us use the describe function to generate key statistics of our dataset, as this will give us a sense of our data, as shown in *Figure 4.4*.

	GPA	Test	Assignment
count	25000.000000	25000.000000	25000.000000
mean	3.005058	9.998700	9.985768
std	1.156037	2.887467	2.894145
min	1.000000	5.000000	5.000000
25%	2.020000	7.500000	7.500000
50%	3.000000	10.000000	10.000000
75%	4.010000	12.500000	12.500000
max	5.000000	15.000000	15.000000

Figure 4.4 – The summary statistics of the numerical columns

From *Figure 4.4*, we can see that the mean GPA is 3.00, the lowest GPA is 1.00, and the highest GPA is 5.00. Both the Test and Assignment columns have a minimum score of 5 and a maximum score of 15. However, we have no sense of the distribution of our target column, as it is a categorical column; we will fix that shortly.

4. Now, let us make a histogram plot of our categorical target variable:

```
plt.hist(df['Graduated'])
plt.show()
```

Running this code produces the plot shown in *Figure 4.5*. Here, we use matplotlib to plot the Graduated column, and we can see almost 17,500 students who successfully graduated and roughly 7,500 students who failed to graduate. Of course, it is only logical to expect a larger number of students will graduate. In ML terms, we have on our plate an unbalanced dataset. However, the good part is that we still have enough samples to train our model from the minority class. Anyway, don't take my word for it; shortly, we will train our models after we complete the data preparation steps.

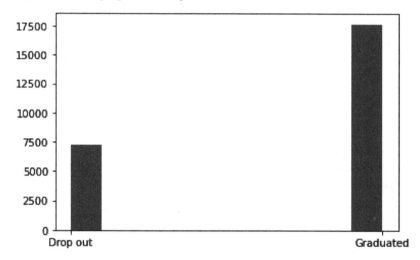

Figure 4.5 – Summary statistics of the numeric columns

5. There are more plots in our notebook to explore, but we will keep it simple, since our main goal in this book is to focus on building models with TensorFlow. However, let's look at one of the very important plots:

```
sns.set(style="darkgrid")
tdc =sns.scatterplot(x ='Library', y ='GPA',
    data = df, hue ='Graduated')
tdc.legend(loc='center left',
    bbox_to_anchor=(1.0, 0.5), ncol=1)
```

Here, we create a scatterplot using seaborn, showing the Library column on the *x* axis and GPA on the *y* axis, and we use the Graduate column to color our data points, as shown in *Figure 4.6*.

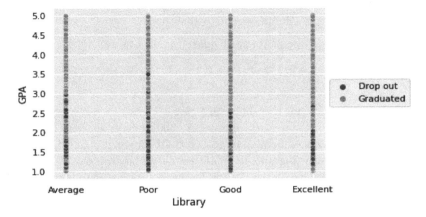

Figure 4.6 – Library versus GPA

From this plot, we can see that there is a good number of students with a GPA above 3.50 who graduated. However, don't assume that everyone above a 3.50 GPA in the Average, Good, and Excellent columns graduated. In fact, let's check this:

```
#To get the number of students with gpa equal to or greater than
3.5 and did not graduate
len(df[(df['GPA']>=3.50)&(df['Graduated']=="Drop out")])
```

When we run this code, we get the total number of students who have a GPA of 3.50 or over and dropped out. In total, we have 76 students who dropped out. Remember, our plot covers 25,000 data points, so don't be surprised if you did not find these data points in the plot in *Figure 4.6*.

Now, let us proceed to prepare our data for modeling.

Data preprocessing

In *Chapter 3*, *Linear Regression with TensorFlow*, we emphasized the need to put our data in the right form, handle missing data, drop irrelevant features, convert categorical values to numeric values, and so on. We will continue in that spirit here:

1. Let us start by converting our labels to numerical values:

    ```
    #Replace the classes in the graduate column
    df['Graduated'] = df['Graduated'].replace(
        ['Graduated', 'Drop out'],[1,0])
    ```

    Here, we assign a value of 1 to students who graduated and a value of 0 to students who dropped out.

2. Now, let us examine the correlation between our numerical data and our target variable using the `corr()` function:

```
df.corr()
```

When we run the code, we get the correlation table shown in *Figure 4.7*.

	GPA	Test	Assignment	Graduated
GPA	1.000000	-0.001451	0.000369	0.652439
Test	-0.001451	1.000000	0.000103	0.244240
Assignment	0.000369	0.000103	1.000000	0.245956
Graduated	0.652439	0.244240	0.245956	1.000000

Figure 4.7 – Correlation table for our dataset

From the highlighted column, we can see that GPA has the strongest correlation with the Graduated column.

3. Now, let us convert our categorical variables to numerical values. We will stick to using dummy variables to one-hot encode our categorical variables:

```
#Converting categorical variables to numeric values
df = pd.get_dummies(df, drop_first=True)
df.head()
```

Here, we drop the first column to avoid the dummy variable trap. When we run the code, we get a new DataFrame, as shown in *Figure 4.8*.

	GPA	Test	Assignment	Graduated	Library_Excellent	Library_Good	Library_Poor	Resources_Offline	Resources_Online	Finance_Unpaid
0	2.07	5.1	9.1	0	0	0	0	0	1	0
1	1.96	10.9	7.8	0	0	0	1	0	1	1
2	3.77	9.3	6.6	1	0	1	0	0	1	0
3	1.82	14.6	6.2	0	0	1	0	1	0	0
4	3.22	14.6	10.9	1	0	0	1	0	0	0

Figure 4.8 – A DataFrame after one-hot encoding

4. Now that we have the attributes in numerical form, let's see how correlated they are with our target variable:

```
tagret_corr= df.corr()
tagret_corr
tagret_corr['Graduated'].sort_values(ascending=False)
```

Once we run the code, it returns the correlation of all the columns with the target variable, as shown in *Figure 4.9*. Our initial numerical variables are still the leading correlation values.

```
Graduated                   1.000000
GPA                         0.652439
Assignment                  0.245956
Test                        0.244240
Study Group_Tier 3          0.114223
Study Time_Excellent        0.076859
Library_Excellent           0.070313
Finance_Unpaid              0.047360
Library_Good                0.028237
Study Time_Good             0.024597
Resources_Offline           0.001883
Study Group_Tier 2         -0.004802
Resources_Online           -0.049768
Scholarships_Yes           -0.059530
Study Time_Poor            -0.071483
Library_Poor               -0.077564
Name: Graduated, dtype: float64
```

Figure 4.9 – A correlation of the attributes with the target column

5. We have successfully converted our data into numerical values, so let us proceed to split the data into attributes (X) and a target (y):

```
# We split the attributes and labels into X and y variables
X = df.drop("Graduated", axis=1)
y = df["Graduated"]
```

6. Don't forget that we need to normalize our data. So, we bring all the attributes to scale for our modeling process:

```
# create a scaler object
scaler = MinMaxScaler()
# fit and transform the data
X_norm = pd.DataFrame(scaler.fit_transform(X),
    columns=X.columns)

X_norm.head()
```

Again, we use `MinMaxScaler` from the scikit-learn library, after which we split our data into training and testing sets. For training, we use 80 percent of our data, and we keep 20 percent as our holdout data to test our model's generalization capability. We set a random state to 10 to ensure that we can reproduce the same data split:

```
# Create training and test sets
#We set the random state to ensure reproducibility

X_train, X_test, y_train, y_test =   train_test_split(
    X_norm, test_size=0.2, random_state=10)
```

Now we are done with data preparation, let us proceed to model building with TensorFlow.

Model building

To build our model, we will start by creating a neural network architecture; here, we will use the sequential API to define the number of layers we want to connect sequentially. As shown in *Figure 4.10*, we have only the input and output layers. Unlike in *Chapter 3, Linear Regression with TensorFlow*, where we predicted numeric values, our output layer here has only one neuron because we are dealing with a binary classification problem. For the output layer, the activation function used depends on the task at hand. When we deal with a binary classification task, we typically use the *sigmoid activation function*; for multi-class classification problems, we commonly use the *softmax activation function*; and when dealing with multi-label classification, we commonly use sigmoid as our activation function.

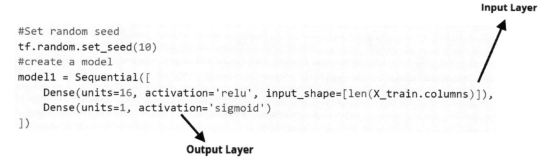

```
#Set random seed
tf.random.set_seed(10)
#create a model
model1 = Sequential([
    Dense(units=16, activation='relu', input_shape=[len(X_train.columns)]),
    Dense(units=1, activation='sigmoid')
])
```

Input Layer

Output Layer

Figure 4.10 – Creating a classification model in TensorFlow

Let us proceed to compile our model. We will use *binary cross entropy* for our loss function when we deal with binary classification and *categorical cross entropy* or *sparse categorical cross entropy* when we deal with multi-class classification problems. In *Chapter 5, Image Classification with Neural Networks*, our discussion will deep-dive into activation functions and more, as we continue to build our understanding and application of neural networks:

```
#compile the model
model1.compile(loss='binary_crossentropy',
    optimizer='adam', metrics='accuracy')
```

Next, let us compile our model. Here, we will use accuracy as our evaluation metric. We will also look at other classification metrics, which we discussed earlier when we begin evaluating our model's performance on test data. After we compile our model, the next step is to fit our model. In *Chapter 1, Introduction to Machine Learning*, we talked about training, validation, and test splits. Since we will deal with a much larger dataset, let's use a validation set to assess our model's performance at the end of each epoch, allowing us to monitor how our model performs on unseen data before we test it out on the hold-out test set. We set our `validation_split` argument to `0.2`; this signifies that we will use 20 percent of our training data for validation during the training process, which will run for 40 epochs:

```
#fit the model
history1= model1.fit(X_train, y_train, epochs=40,
    validation_split=0.2)
```

In *Figure 4.11*, we can see the last five epochs of our model's output:

```
Epoch 46/50
500/500 [==============================] - 1s 3ms/step - loss: 0.0189 - accuracy: 0.9955 - val_loss: 0.0189 - val_accuracy: 0.9955
Epoch 47/50
500/500 [==============================] - 1s 2ms/step - loss: 0.0187 - accuracy: 0.9964 - val_loss: 0.0190 - val_accuracy: 0.9950
Epoch 48/50
500/500 [==============================] - 2s 3ms/step - loss: 0.0187 - accuracy: 0.9954 - val_loss: 0.0183 - val_accuracy: 0.9977
Epoch 49/50
500/500 [==============================] - 1s 3ms/step - loss: 0.0187 - accuracy: 0.9952 - val_loss: 0.0176 - val_accuracy: 0.9967
Epoch 50/50
500/500 [==============================] - 1s 2ms/step - loss: 0.0179 - accuracy: 0.9956 - val_loss: 0.0189 - val_accuracy: 0.9952
<keras.callbacks.History at 0x7dfe06f03460>
```

Figure 4.11 – Model training (the last five epochs)

The model reaches a training accuracy of 99.35% and a validation accuracy of 99.33%. Using just two layers and three simple steps in less than five minutes, we have arrived at almost 100 percent accuracy on both training and validation data. These results are impressive, yes; however, it is important to know this isn't always the case, especially when we work with more complex datasets. They may require more complex architectures and longer training times to achieve good results. We will see this in *Section 2* of this book, where we will work with images. Before we proceed to evaluate our model, let us look at our model's architecture using the `summary` function:

```
model1.summary()
```

When we run this line of code, we generate the model's architecture:

```
Model: "sequential"

_____
 Layer (type)                Output Shape              Param #
=================================================================
 dense (Dense)               (None, 16)                256

 dense_1 (Dense)             (None, 1)                 17


=================================================================
Total params: 273
Trainable params: 273
Non-trainable params: 0
_____
```

The output shape shows us that the first dense layer (input layer) has 16 neurons and 256 params, since we passed in 16 attributes (16 columns x 16 neurons = 256 params), while the dense_1 layer (the output layer) has 1 neuron and 17 params (17 columns x 1 neuron = 17 params). The total params are 273 and all the params are trainable, so we have 273 here, which means there will be zero non-trainable params. Now that we are done with model building, let's shift our attention to evaluating our model. How well will it perform on the test data?

Classification performance evaluation

To evaluate our model in TensorFlow, all we need is one line of code – using the evaluate function on our model:

```
# Evaluate the Classication model
eval_model=model1.evaluate(X_test, y_test)
eval_model
```

We then generate our model's performance on our holdout data:

```
157/157 [==============================] - 1s 4ms/step - loss: 0.0592
- accuracy: 0.9944
[0.05915425345301628, 0.9944000244140625]
```

We arrive at an accuracy of 99.44% on our test data. This is good; however, let us look at the other classification metrics we talked about earlier in this chapter:

```
y_pred=model1.predict(X_test).flatten()
y_pred = np.round(y_pred).astype('int')
df_predictions = pd.DataFrame(
    {'Ground_Truth': y_test, 'Model_prediction': y_pred},
```

```
    columns=[ 'Ground_Truth', 'Model_prediction'])
len(df_predictions[(df_predictions[
    'Ground_Truth']!=df_predictions['Model_prediction'])])
```

We generate our model's prediction on the test data. Then, we convert the probabilities using the np.round() function and convert the data type to integers. Then, we create a pandas DataFrame, after which we generate the number of the misclassified labels in our DataFrame. In our case, the model misclassified 28 out of 5,000 data points in our test set. Now, we will generate a confusion matrix and classification reports to evaluate our model:

```
#Generating the confusion matrix
eval = confusion_matrix(y_test, y_pred)
print(eval)
```

Running this code generates the confusion matrix shown in *Figure 4.12*.

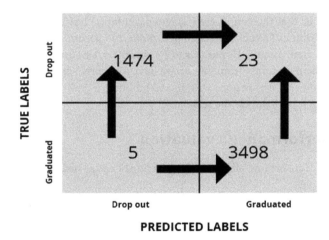

Figure 4.12 – A confusion matrix for our student dropout model

The horizontal arrows point in the direction of the true values, while the vertical arrows are in the direction of the predicted labels. Our ground truth had (5 + 3,498) = 3,503 students in the graduated class, and our model predicted (3498 + 23) = 3,521 in our graduate class. Meanwhile, in the dropout class, our model predicted (5 + 1474) = 1,479 students dropped out, as against the ground truth (1,474 + 23) = 1497. From *Figure 4.14*, we can see that the model wrongly predicted that 23 students who dropped out were graduates and 5 students who graduated were dropouts.

Next, let us print out our classification report:

```
class_names = [ 'Drop Out', 'Graduated']
print(classification_report(y_test, y_pred,
    target_names=class_names))
```

Now that we have printed out our classification report, as shown in *Figure 4.15*, we can see our model's precision, recall, and F1 score across both classes in our dataset, as well as the macro and weighted averages.

	precision	recall	f1-score	support
Drop Out	1.00	0.98	0.99	1497
Graduated	0.99	1.00	1.00	3503
accuracy			0.99	5000
macro avg	1.00	0.99	0.99	5000
weighted avg	0.99	0.99	0.99	5000

Figure 4.13 – The classification report

From the highlighted details in *Figure 4.13*, we can see the model's precision, recall, and F1 score. We have come a long way; this is a good result. If we wish to improve our result, we can try more experiments. Also, error analysis is very useful to help us understand misclassified data. We can drill down into the misclassified students, trying to understand patterns or common characteristics in cases the model failed to predict correctly. This can lead to further insights or help us identify issues regarding our data quality, among other possibilities. However, we will not delve into error analysis here. You have done a good job and achieved good results in both classes.

Now, let's save the model and present it to the manager using the `save` function:

```
#saving our model
model1.save('classification_model.h5')
```

We have completed our task, and we have a near-perfect model.

Now, you should be able to build a real-world classifier with TensorFlow for structured data problems, using what you learned from our case study in this chapter.

Summary

In this chapter, we discussed classification modeling and looked at the main types of classification problems. We also discussed the main types of metrics for the evaluation of classification models and how to best apply them to real-world use cases. Then, we looked at a real-world use case, where we learned how to build, compile, and train a classification model with TensorFlow for a binary classification problem.

Finally, we learned, hands-on, how to evaluate our classification models. We have now completed the first section of this book. Get ready for the next sections, where we will see the power of TensorFlow in its full glory as we work on unstructured data (image and text data).

Questions

Let's test what we learned in this chapter.

1. What is classification modeling?

2. What is the difference between multi-class and multi-label classification problems?

3. You work for a streaming company offering interesting children content. Which metrics, between precision and recall, will you focus on improving, and why?

4. Your company is building a loan prediction system to offer loans to clients. Which metrics, between precision and recall, will you focus on improving, and why?

Further reading

To learn more, you can check out the following resources:

*   Amr, T., 2020. *Hands-On Machine Learning with scikit-learn and Scientific Python Toolkits.* [S.l.]: Packt Publishing.

*   Beger, A., 2016. *Precision-Recall Curves. SSRN Electronic Journal.*

*   Raschka, S. and Mirjalili, V., 2019. *Python Machine Learning – Third Edition.* Packt Publishing.

*   *TensorFlow guide*: https://www.TensorFlow.org/guide.

Part 2 – Image Classification with TensorFlow

In this part, you will learn to build both binary and multiclass image classifiers with **convolutional neural networks** (**CNNs**), understand how to improve the model's performance by tuning the hyperparameters, and how to handle the problem of overfitting. By the end, you should be comfortable with building real world image classifiers using transfer learning.

This section comprises the following chapters:

- *Chapter 5, Image Classification With Neural Networks*
- *Chapter 6, Improving the Model*
- *Chapter 7, Image Classification with Convolutional Neural Networks*
- *Chapter 8, Handling Overfitting*
- *Chapter 9, Transfer Learning*

5

Image Classification with Neural Networks

Up until this point, we have built models to solve both regression and classification problems on structured data with much success. The next question that comes to mind is: can we build models that can tell the difference between a dog and a cat, or a car and a plane? Today, with the aid of frameworks such as **TensorFlow** and **PyTorch**, developers can now build such ML solutions with a few lines of code.

In this chapter, we will explore the anatomy of **neural networks** and learn how we can apply them to building models for computer vision problems. We will start by examining what a neural network is and the architecture of a multilayer neural network. We will look at some important ideas such as forward propagation, backward propagation, optimizers, loss function, learning rate, and activation functions, and where and how they fit in.

After we build a solid base in the core fundamentals, we will build an image classifier using a custom dataset from TensorFlow. Here, we will walk through the end-to-end process of model building using the TensorFlow dataset. The good part of using these custom datasets is that the bulk of the preprocessing steps are already done, and our data can be modeled without any blockers. So, we will use this dataset to build a neural network with a few lines of code in TensorFlow with the **Keras** API, so that our model will be able to tell the difference between a bag and a shirt, and a shoe and a coat.

In this chapter, we'll cover the following topics:

- The anatomy of neural networks
- Building an image classifier with a neural network

Technical requirements

We will be using **Google Colab** to run the coding exercise that requires `python >= 3.8.0`, along with the following packages that can be installed using the `pip install` command:

- `tensorflow>=2.7.0`
- `tensorflow-datasets==4.4.0`
- `pillow==8.4.0`
- `pandas==1.3.4`
- `numpy==1.21.4`
- `matplotlib >=3.4.0`

The code for this chapter is available at `https://github.com/PacktPublishing/TensorFlow-Developer-Certificate-Guide/tree/main/Chapter%205`. Also, solutions to all exercises can be found in the GitHub repository itself.

The anatomy of neural networks

In the first section of this book, we talked about models. These models that we spoke about and used for various use cases are neural networks. A neural network is a deep learning algorithm inspired by the functionality of the human brain, but by no means does it operate like the human brain. It learns useful representation of the input data using a layered approach, as shown in *Figure 5.1*:

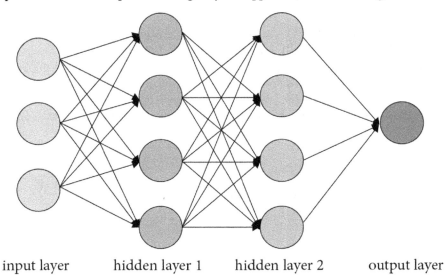

input layer   hidden layer 1   hidden layer 2   output layer

Figure 5.1 – Neural network

Neural networks are ideal for tackling complex problems due to their ability to identify very complex patterns in data. This makes them well suited for building solutions around text and image data (unstructured data), tasks that traditional machine learning algorithms struggle with. Neural networks develop rules to map input data to the target or labels using layered representation. When we train them on labeled data, they learn the patterns and use this knowledge to map the new input data to their corresponding labels.

In *Figure 5.1*, we see all the neurons of the input layer are connected to the neurons of the first hidden layer, and all the neurons of the first hidden layer are connected to all the neurons of the second hidden layer. The same applies from the second hidden layer to the outer layer. This type of network, where each layer's neurons are fully connected to the neurons of the next layer, is called a **fully connected neural network**. A neural network with more than two hidden layers is called a **deep neural network** (**DNN**) and the depth of the network is determined by its number of layers.

Let's take a deep dive into the individual layers of a neural network architecture:

- **Input layer**: This is the layer through which we fed the input data (text, image, tabular data) into the network. Here, we have to specify the right input shape, something we have done previously in our regression case study in *Chapter 3, Linear Regression With TensorFlow*, and in our classification case study in *Chapter 4, Classification With TensorFlow*. It is important to note that the input data will be presented to our neural network in numerical format. In this layer, no computation takes place. It's more of a passthrough layer to the hidden layer.

- **Hidden layer**: This is the next layer and it lies between the input and output layers. It is referred to as hidden because it is not visible to external systems. Here, lots of computation takes place to extract patterns from our input data. The more layers we add to the hidden layer, the more complex our model becomes and the more time it takes to process our data.

- **Output layer**: This layer produces the output of the neural network. The number of output layer neurons is determined by the task at hand. If we have a binary classification task, we will use one output neuron, while for multiclass classification, such as in our case study where we had 10 different labels, we will have 10 neurons, one for each of the classes in our data.

We now know the layers of a neural network, but the key questions are: how does a neural network work, and what enables it to take a special position in machine learning?

Neural networks solve complex tasks by the application of both forward and backward propagation. Let's start by examining forward propagation.

Forward propagation

Imagine we want to teach our neural network to effectively identify the images shown in *Figure 5.2*. We will pass lots of representative samples of each of the images we want our neural network to recognize. The idea here is that our neural network will learn from the samples and use what it has learned to identify new items within the sample space. Let's say, for example, we want our model to recognize shirts; we will pass shirts of different colors and sizes. Our model will learn what defines a shirt, irrespective of its color, size, or style. This learned representation of the core attributes of a shirt is what the model will use to identify new shirts.

Figure 5.2 – Sample images from Fashion MNIST dataset

Let us look at what happens under the hood. In our training data, we pass the images (X) through our model $f(x) .. \rightarrow \hat{y}$, where $\hat{y}$ is the model's predicted output. Here, the neural network randomly initializes weights that are used to predict the output ($\hat{y}$). This process is called **forward propagation** or **forward pass** and is depicted in *Figure 5.3*.

> **Note**
>
> Weights are trainable parameters that are updated during the training process. After training, the model's weights are optimized to the specific dataset it is trained on. If we tune the weight properly during training, we can develop a well-performing model.

As input data flows through the network, it experiences transformations due to the impact of the node's weight and bias, as shown in *Figure 5.3*, thus producing a new set of information that will now pass through an *activation function*. If the new information learned is desired, the activation function triggers an output signal that serves as input to the next layer. This process continues until an output is generated in the output layer:

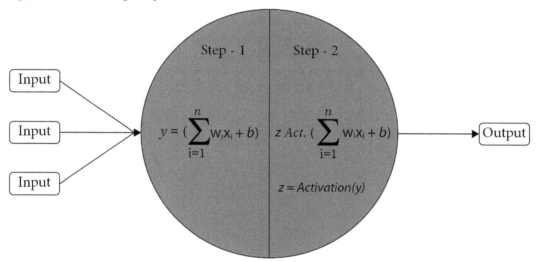

Figure 5.3 – Forward propagation of a neural network

Let's talk a bit more about activation functions and what they do in our neural network.

Activation functions

Imagine you had to pick the good apples from a basket of apples. By inspecting the apples, you can pick the good ones and drop the bad ones. This is how an activation function works – it plays the role of a separator and thus defines what will pass through, which in our case is the useful representation it has learned, and drops the non-useful data. In essence, it helps to extract useful information, such as the good apples, and drop the useless data, which in our scenario are the bad apples. Now, the activation function determines which connected neuron of the next layer will be activated. It uses mathematical operations to determine whether a learned representation is useful enough for the next layer or not.

Activation functions can add nonlinearity to our neural network, a characteristic that is required for neural networks to learn complex patterns. There are different activation functions; for output layers, the selection of activation function depends on the type of task at hand:

- For binary classification, we usually use sigmoid function because it maps the input to output values between 0 and 1, representing the probability of belonging to a particular class. We usually set the threshold point as 0.5, hence values above this point are set to 1 and values below it is set to 0.

- For multiclass classification, we use **softmax activation** as the output layer's activation function. Let's say we want to build an image classifier to classify four fruits (apples, grapes, mangoes, and oranges) as illustrated in *Figure 5.4*. One neuron is assigned in the output layer to each of the fruits and we will apply the softmax activation function to generate the likelihood of the output being one of the fruits we want to predict. When we sum up the probabilities of it being an apple, grape, mango, and orange, we get 1. For classification, we select the class with highest probability of the four fruits as the output label from the probabilities generated by the Softmax activation function. In this case, the output with the highest probability is an orange:

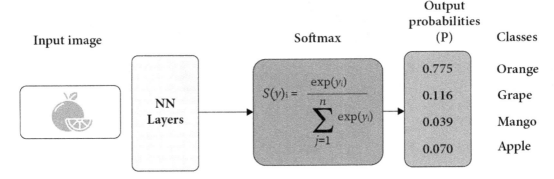

Figure 5.4 – Application of the SoftMax activation function

For hidden layers, we will use the **rectified linear unit (ReLU)** activation function. This activation function removes negative values (useless representations), while it passes learned representations with values greater than 0. ReLU offers excellent performance for hidden layers as it converges quickly as well as supports backward propagation, a concept we will be discussing next.

> **Note**
>
> It is more efficient to use sigmoid for binary classification, when we do this, we have one output neuron as against two output neurons which would be the case when we use Softmax. Also, it is easier to understand that we are working on a case of binary classification when we read the code.

Backward propagation

When we begin training a model, the weights are initially random, making it more likely that the model will guess wrongly that the fruit in *Figure 5.4* is an orange. Here comes the intelligence of our neural network; it autocorrects itself, as shown in *Figure 5.5*:

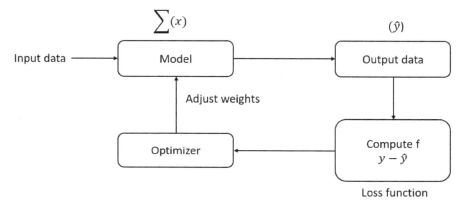

Figure 5.5 – Forward and backward propagations of a neural network

Here, the neural network measures how correct the predicted output ($\hat{y}$) is in comparison to the ground truth (y). This loss is computed by the **loss function**, which can also be referred to as the **cost function**. This information is passed on to an **optimizer**, whose job is to update the weights of the layers in the neural network with the aim of reducing the loss over the next iterations, thus getting our prediction closer to the ground truth. This process continues until we achieve **convergence**. Convergence occurs when the model is trained such that the loss is at its barest minimum.

The loss function is applied with respect to the task at hand. When we are working on a binary classification task, we use **binary cross-entropy**; for multiclass classification, if the target labels are integer values (for example, 0 to 9) we use **sparse categorical cross-entropy**, whereas we use **categorical cross-entropy** if we decide to one-hot encode our target labels. Like loss functions, we also have different types of optimizers; however, we will experiment with **stochastic gradient descent** (**SGD**) and the **Adam optimizer**, which is an improved version of SGD. Hence, we will use this as our default optimizer.

Learning rate

We now know that weights are randomly initialized, and optimizers aim to use this information about the loss function to update the weights with a view to achieving convergence. Neural networks use optimizers to iteratively update the weights until the loss function is at a minimum, as shown in *Figure 5.6*. Optimizers let you set an important hyperparameter called the **learning rate**, which controls the speed of convergence and is how our model learns. To get to the bottom of the slope, we will have to take steps toward the base (see *Figure 5.6*):

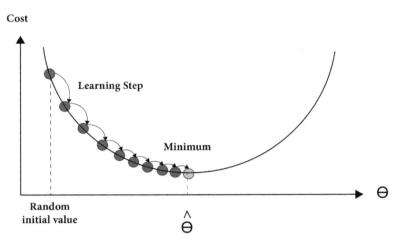

Figure 5.6 – Gradient descent

The step size we take will determine how quickly we get to the base. If we take very small steps, it will take too long to reach the base and lead to slower convergence, and there is also a risk that the optimization process could get stuck along the way to the minimum point. On the flip side, if the steps are too large, there is a risk we may overshoot the minimum and experience erratic and unstable training behavior. The right step size will get us to the base of the slope in time without overshooting the minimum point. This step size we refer to here is the learning rate.

We have now covered the intuition behind neural networks at a high level. Let us proceed and look at our case study, directly applying what we have just learned.

Building an image classifier with a neural network

We are back at our fictional company, and we want to use the intuition of neural networks to build an image classifier. Here, we are to teach computers to identify clothing. Thankfully, we do not need to find data in the wild; we have TensorFlow datasets that include the fashion dataset. In our case study, our aim is to classify a fashion dataset made up of 28 x 28 grayscale images into 10 classes (from 0 to 9) with pixel values between 0 and 255, using a well-known dataset called the *Fashion MNIST dataset*. This dataset is made up of 60,000 training images and 10,000 test images. Our dataset has all the images in the same shape, so we have little preprocessing to do. The idea here is for us to build a neural network quickly with little preprocessing complexities.

To train the neural network, we will pass the training images with the idea that our neural network will learn to map the images (X) to their corresponding labels (y). After we have concluded the training process, we will use our test set to evaluate the model on new unseen images. Again, the idea is that the model will correctly identify test images based on what it has learned throughout the training process. Let's begin.

Loading the data

Here, we will start with learning how to work with images using TensorFlow datasets. In *Chapter 7, Image Classification with Convolutional Neural Networks*, we will work on real-world images that will require more work to model our data; however, it will build on what we will learn here. That said, let us see how we can load our custom dataset from TensorFlow:

1. Before we can load our data, we need to load the necessary libraries. Let's do that here:

```
import tensorflow as tf
from tensorflow import keras
import pandas as pd
import random
import numpy as np
import matplotlib.pyplot as plt #helper libraries
from tensorflow.keras.utils import plot_model
```

2. Next, we import the `fashion_mnist` dataset from TensorFlow and create our training and testing dataset using the `load_data()` method:

```
#Lets import the fashion mnist
fashion_data = keras.datasets.fashion_mnist
#Lets create of numpy array of training and testing data
(train_images, train_labels), (test_images,
    test_labels) = fashion_data.load_data()
```

If everything goes according to plan, we should get an output as shown in *Figure 5.7*:

```
Downloading data from https://storage.googleapis.com/tensorflow/tf-keras-datasets/train-labels-idx1-ubyte.gz
32768/29515 [==============================] - 0s 0us/step
40960/29515 [==============================] - 0s 0us/step
Downloading data from https://storage.googleapis.com/tensorflow/tf-keras-datasets/train-images-idx3-ubyte.gz
26427392/26421880 [============================] - 1s 0us/step
26435584/26421880 [============================] - 1s 0us/step
Downloading data from https://storage.googleapis.com/tensorflow/tf-keras-datasets/t10k-labels-idx1-ubyte.gz
16384/5148 [=================================================================================] - 0s 0us/step
Downloading data from https://storage.googleapis.com/tensorflow/tf-keras-datasets/t10k-images-idx3-ubyte.gz
4423680/4422102 [===========================] - 0s 0us/step
4431872/4422102 [===========================] - 0s 0us/step
```

Figure 5.7 – Data import from TensorFlow datasets

3. Now, rather than using numeric labels, let us create labels that match our data such that we can call a dress a dress, rather than call it number 3. We will do that by creating a list of our labels, which we will map to the corresponding numeric values:

```
#We create a list of the categories
class_names=['Top', 'Trouser','Pullover', 'Dress', 'Coat',
    'Sandal', 'Shirt', 'Sneaker', 'Bag', 'Ankleboot']
```

Now that we have our data, let us explore the data and see what we can find. Rather than agree with everything we are told, let's explore the data to verify the size, the shape, and the data distributions.

Performing exploratory data analysis

After we load the data, the next step is to examine it to get a sense of what the data is. Of course, in this instance, we have some basic information from TensorFlow about the data distribution. Also, we have the data already available in training and test sets. However, let us confirm all the details using code, as well as view the class distribution of our target label:

1. We will use the `matplotlib` library to generate image samples at index `i`, where `i` falls within the 60,000 training samples:

    ```
    # Display a sample image from the training data (index 7)
    plt.imshow(train_images[7])
    plt.grid(False)
    plt.axis('off')
    plt.show()
    ```

    We run the code using index 7, which returns a top as seen in *Figure 5.7*:

    Figure 5.8 – A photo of a pullover at index 7 of the Fashion MNIST dataset

    We can switch the index values to see other apparels within the dataset; however, that is not the goal here. So, let's proceed with our exploratory data analysis.

2. Let's look at the sample of our data:

    ```
    #Lets check the shape of our training images and testing images
    train_images.shape, test_images.shape
    ```

    As expected, we can see the training images consist of 60,000 28 x 28 images, and the test images are 10,000 in number and 28 x 28 in resolution:

    ```
    ((60000, 28, 28), (10000, 28, 28))
    ```

3. Next, let us check the distribution of the data. It's best practice to see how your data is distributed to ensure there is enough representation for each class of clothing we want to train the model on. Let's do that here:

    ```
    df=pd.DataFrame(np.unique(train_labels,
        return_counts=True)).T
    dict = {0: <Label>,1: <Count>}
    ```

```
df.rename(columns=dict,
    inplace=True)
df
```

This returns a `DataFrame` as shown in *Figure 5.9*. We can see that all the labels have the same number of samples:

	Label	Count
0	0	6000
1	1	6000
2	2	6000
3	3	6000
4	4	6000
5	5	6000
6	6	6000
7	7	6000
8	8	6000
9	9	6000

Figure 5.9 – DataFrame showing labels and their counts

Of course, this type of data is more likely to be found in a controlled setting such as academia.

4. Let us visualize some sample images from the training data here. Let's look at 16 samples from our training data:

```
plt.figure(figsize=(9,9))
for i in range(16):
    plt.subplot(4,4,i+1)
    plt.xticks([])
    plt.yticks([])
    plt.grid(False)
    plt.imshow(train_images[i])
    plt.title(class_names[train_labels[i]])
plt.show()
```

5. When we run the code, we get the image in *Figure 5.10*:

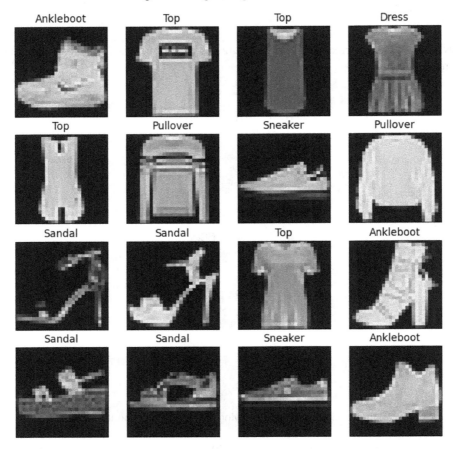

Figure 5.10 – 16 randomly selected images from the Fashion MNIST dataset

Now we have confirmed the data size, the data distribution, and the shape, and seen some sample images and labels. Before we proceed with building and training our image classifier, recall our data is made up of grayscale images with values from 0 to 255. To bring the data to scale, we will have to normalize the data to improve the performance of our model during training. We can do this by simply dividing the training and testing data by 255:

```
#it's important that the training and testing set are preprocessed in
the same way.
train_images=train_images/255.0
test_images=test_images/255.0
```

Now that we have normalized our data, we are all set for modeling it. Let's proceed with building our image classifier next.

Building the model

Let us put everything we have learned so far in this chapter into action:

```
#Step 1:  Model configuration
model=keras.Sequential([
    keras.layers.Flatten(input_shape=(28,28)),
    keras.layers.Dense(64, activation="relu"),
    keras.layers.Dense(10,activation="softMax")


])


#Here we flatten the data
```

The code we use to build our model is similar to what we used in *Part 1* of this book. We start by creating a sequential model using the Sequential API to define the number of layers we want to connect sequentially. If you are a keen observer, you will notice our first layer is a flatten layer. This is used to flatten the image data into a 1D array that will be passed into the hidden layer. The input layer has no neurons; it works as a data preprocessing layer, presenting the hidden layer with data flattened into a 1D array.

Next, we have one hidden layer of 64 neurons, and we apply a ReLU activation function to this hidden layer. Finally, we have an output layer of 10 neurons – one neuron for each output. We use a softmax function since we are working on multiclass classification. Softmax returns results in the form of probabilities across all classes. If you recall from the *Activation functions* section, the sum of the output probabilities adds up to 1, and the output with the largest probability value is the predicted label.

Now that we are done with model building, let us proceed with compiling our model.

Compiling the model

The next step is to compile the model. We will use the `compile` method to do so. Here, we pass in the optimizer we wish to use; in this case, we apply **Adam**, which is our default optimizer. We also specify the loss and the evaluation metrics. We use sparse categorical cross-entropy for our loss since our labels are numeric values. For our evaluation metrics, we use accuracy, since our dataset is balanced. The accuracy metric will give a true reflection of our model's performance:

```
#Step 2: Compiling the model, we add the loss, optimizer and
evaluation metrics here
model.compile(optimizer='adam',
    loss=>sparse_categorical_crossentropy',
    metrics=[<accuracy>])
```

Before we proceed to fitting our model, let's look at some ways of visualizing our model and its parameters.

Model visualization

To visualize our model, we use the `summary()` method. This provides us with a detailed visual representation of the model's architecture, the layers, the number of parameters (trainable and non-trainable), and the output shape:

```
model.summary()
```

When we run the code, it returns the model's details as illustrated in *Figure 5.11*:

```
Model: "sequential"
```

Layer (type)	Output Shape	Param #
flatten (Flatten)	(None, 784)	0
dense (Dense)	(None, 64)	50240
dense_1 (Dense)	(None, 10)	650

```
Total params: 50,890
Trainable params: 50,890
Non-trainable params: 0
```

Figure 5.11 – Model summary

From *Figure 5.11*, we can see that the input layer has no parameters but an output shape of 784, which is the result of flattening our 28 × 28 image to a 1D array. To get the number of parameters of the dense layer, it's 784 × 64 + 64 = 50240 (recall , where X is the input data, w is the weights, and b is the bias). The output layer (`dense_1`) has a shape of 10, with one neuron representing each class and 650 parameters. Recall the output from one layer serves as the input to the next layer. So, 64 × 10 + 10 = 650, where 64 is the output shape of the hidden layer and the input shape of the output layer.

On the other hand, we can also display our model as a flowchart, as seen in *Figure 5.12*, by using the following code:

```
plot_model(model, to_file='model_plot.png', show_shapes=True,
    show_layer_names=True)
```

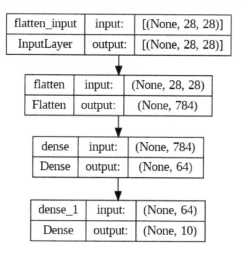

Figure 5.12 – Model's flowchart

This also gives us a sense of our model's structure. The plot we generated will be saved with the filename `model_plot.png`. Here, we set `show_shapes` to `true`; this will display the output shapes of each layer in the plot. We also set the `show_layer_name` to `true` to show the names of the layers in the plot, as illustrated in *Figure 5.12*.

Next, let us fit our model to the training data.

Model fitting

By now, you should be familiar with this process. With a single line of code, we can use the `fit` method to fit our training images (X) and training labels (y):

```
#Step 3: We fit our data to the model
  history= model.fit(train_images, train_labels, epochs=5)
```

Here, we fit the data for five epochs. Our model returns the loss and accuracy:

```
1875/1875 [==============================] - 4s 2ms/step - loss:
0.5206 - accuracy: 0.8183
Epoch 2/5
1875/1875 [==============================] - 4s 2ms/step - loss:
0.3937 - accuracy: 0.8586
Epoch 3/5
1875/1875 [==============================] - 4s 2ms/step - loss:
0.3540 - accuracy: 0.8722
Epoch 4/5
1875/1875 [==============================] - 4s 2ms/step - loss:
0.3301 - accuracy: 0.8790
```

```
Epoch 5/5
1875/1875 [==============================] - 4s 2ms/step - loss:
0.3131 - accuracy: 0.8850
```

We can see that in just five epochs, our model has achieved an accuracy of 0.8850. This is a good start considering we trained our model for a very small number of epochs. Next, let us observe our model's performance during training by plotting the loss and accuracy plots.

Training monitoring

We return a history object when we fit our training data. Here, we use the history object to create a loss and accuracy curve. Here is the code to make the plots:

```
# Plot history for accuracy
plt.plot(history.history['accuracy'])
plt.title('model accuracy')
plt.ylabel('accuracy')
plt.xlabel('epoch')
plt.legend(['Train'], loc='lower right')
plt.show()
# Plot history for loss
plt.plot(history.history['loss'])
plt.title('model loss')
plt.ylabel('loss')
plt.xlabel('epoch')
plt.legend(['Train'], loc='upper right')
plt.show()
```

When we run the code, we get back two plots as shown in *Figure 5.13*. We can see the training accuracy is still rising at the end of the fifth epoch, while the loss is still falling, although the rate is not rapid as it moves closer to 0:

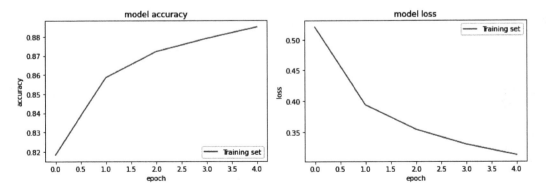

Figure 5.13 – Accuracy and loss plots

Perhaps if we train for longer, we could see an improved performance. In the next chapter, we will examine what happens if we do train for longer, as well as look at other approaches to improve our model's performance. Here, the aim is to understand what the plot means and gain enough information to direct our next line of action. Let's evaluate our model on the test set.

Evaluating the model

We evaluate the overall performance of our model on the test set as follows:

```
test_loss,test_acc =model.evaluate(test_images,test_labels)
print('Test Accuracy: ', test_acc)
```

We get an accuracy of 0.8567 on the test set. The difference between the training accuracy and the test accuracy is a common problem in machine learning that we refer to as **overfitting**. Overfitting is a key issue in machine learning, and we will look at overfitting and various ways of handling it in *Chapter 8, Handling Overfitting*.

Next, let us make some predictions with our trained neural network.

Model prediction

To make predictions on the model, we use the `model.predict()` method on unseen data from our test set. Let's look at what the model predicts on the first instance of our test data:

```
predictions=model.predict(test_images)
predictions[0].round(2)
```

When we run the code, we get back an array of probabilities:

```
array([0.  , 0.  , 0.  , 0.  , 0.  ,
    0.13, 0.  , 0.16, 0.  , 0.7 ],
    dtype=float32)
```

If we inspect the probabilities, we see that the probability is highest at the ninth element. So, there is a 70% chance that this is our label. We will use `np.argmax` to extract the label and compare it to the test label at index 0:

```
np.argmax(predictions[0]),test_labels[0]
```

We see that both the predicted label and test label return a value of 9. Our model got this prediction right. Next, let us plot 16 random images and compare the predicted results with the ground label. This time, rather than returning the numeric values of our labels, we will return the labels themselves for visual clarity:

```
# Let us plot 16 random images and compare the labels with the model's
prediction
figure = plt.figure(figsize=(9, 9))
```

```
for i, index in enumerate(np.random.choice(test_images.shape[0],
size=16, replace=False)):
    ax = figure.add_subplot(4,4,i + 1,xticks=[], yticks=[])
    # Display each image
    ax.imshow(np.squeeze(test_images[index]))
    predict_index = np.argmax(predictions[index])
    true_index = test_labels[index]
    # Set the title for each image
    ax.set_title(f»{class_names[predict_index]} (
{class_names[true_index]})",color=(
        "green" if predict_index == true_index else «red»))
```

The result is shown in *Figure 5.14*. Although the model was able to classify 10 items correctly, it failed on one sample where it classified a shirt as a pullover:

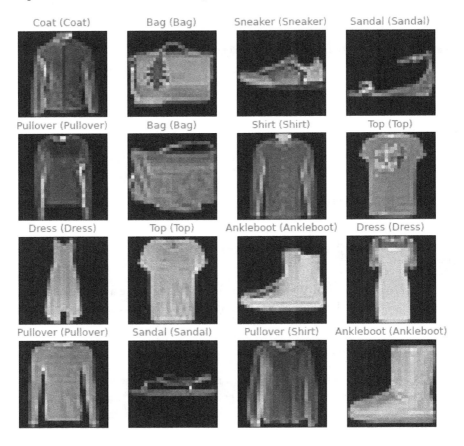

Figure 5.14 – Visualizing the model's prediction on test data

In just a few lines of code, we have trained an image classifier. We reached an accuracy of 88.50% on our training data in five epochs and 85.67% on our test data. It is important to note that this is a toy dataset that is great for learning; however, real-world images are more complex and the training will take much longer and, in many instances, a more complex model architecture will be required.

In this chapter, we have covered a lot of new concepts that will be very useful in later chapters and in the exams as well.

Summary

In this chapter, we discussed image classification modeling. Now, you should be able to explain what a neural network is, as well as forward and backward propagation. You should know the role of loss functions, activation functions, and optimizers in a neural network. Also, you should be able to find your way around loading data from a TensorFlow dataset. Finally, you should be familiar with how to build, compile, fit, and train a neural network for image classification as well as evaluate the model, plot the loss and accuracy curves, and interpret these visualizations.

In the next chapter, we will explore several ideas we can apply to improve our model's performance.

Questions

Let's test what we learned in this chapter:

1. What is the function of the activation function?
2. How does backward propagation work?
3. What is the purpose of the input, hidden, and output layers?
4. Using a TensorFlow dataset, load a handwritten digits dataset after which you will build, compile, train, and evaluate an image classifier. It's a similar exercise to our case study. Go for it.

Further reading

To learn more, you can check out the following resources:

*   Amr, T., 2020. *Hands-On Machine Learning with scikit-learn and Scientific Python Toolkits*. [S.l.]: Packt Publishing.
*   Vasilev, I., 2019. *Advanced Deep Learning with Python*. 1st ed. Packt Publishing.
*   Raschka, S. and Mirjalili, V., 2019. *Python Machine Learning*. 3rd ed. Packt Publishing.
*   Gulli, A., Kapoor, A. and Pal, S., 2019. *Deep Learning with TensorFlow 2 and Keras*. Birmingham: Packt Publishing.
*   *TensorFlow Guide* https://www.TensorFlow.org/guide

6
Improving the Model

The goal of modeling in machine learning is to ensure our model generalizes well on unseen data. Throughout our journey as data professionals who build models with neural networks, we are likely to come across two main issues: underfitting and overfitting. **Underfitting** is a scenario in which our model lacks the necessary complexity to capture underlying patterns in our data, while **overfitting** occurs when our model is too complex such that it not only learns the patterns but also picks up noise and outliers in our training data. In this case, our model performs exceptionally well on training data but fails to generalize well on unseen data. *Chapter 5, Image Classification with Neural Networks,* examined the science behind neural networks. Here, we will explore the art of fine-tuning neural networks to build optimally performing models for image classification. We will explore various network settings in a hands-on fashion to gain an understanding of the impact of each of these settings (hyperparameters) on our model's performance.

Beyond exploring the art of hyperparameter tuning, we will also explore various ways of improving our data quality using data normalization, data augmentation, and the use of synthetic data to improve the generalization capabilities of our model. In the past, there was a lot of emphasis on building complex networks. However, in more recent times, there has been an increased interest in enhancing the performance of neural networks using data-centric strategies. The use of these data-centric strategies does not erode the need for careful model design; rather, we can look at them as complementary strategies working in tandem toward a desired goal, thus enhancing our ability to build optimal models with good generalization capabilities.

In this chapter, we will cover the following topics:

- Data is key
- Fine-tuning hyperparameters of a neural network

By the end of this chapter, you will be equipped to effectively navigate the challenges posed by overfitting and underfitting using a combination of model-centric and data-centric ideas when building models with neural networks.

Technical requirements

We will be using a **Google Colaboratory** (**Google Colab**) notebook as our work environment as it is a free, cloud-based Jupyter Notebook that is easy to use and provides us with GPU and TPU backends. We will be using Google Colab to run the coding exercise that requires `python >= 3.8.0`, along with the following packages that can be installed using the `pip install` command:

- `tensorflow>=2.7.0`
- `tensorflow-datasets==4.4.0`
- `pandas==1.3.4`
- `numpy==1.21.4`

Data is key

When it comes to improving the performance of a neural network, or any other machine learning model for that matter, the importance of good data preparation cannot be overemphasized. In *Chapter 3, Linear Regression with TensorFlow*, we saw the impact that normalizing our data had on the model's performance. Beyond data normalization, there are other data preparation techniques that can make a difference in our modeling process.

As you must have recognized by now, machine learning requires investigating, experimenting, and applying different techniques, depending on the problem at hand. To ensure we have an optimally performing model, our journey should start by looking at our data thoroughly. Do we have enough representative samples from each of the target classes? Is our data balanced? Have we ensured the absence of incorrect labels? Do we have the right type of data? How are we dealing with missing data? These are some of the questions we have to ask and handle before the modeling phase.

Improving the quality of our data is a multi-faceted endeavor involving different techniques, such as engineering new features from existing ones in our data by applying some data preprocessing techniques such as data normalization. When we are working with imbalanced datasets, where we are short of representative samples of the minority class, the logical thing would be to gather more data on the minority class; however, this is not practical in all instances. In such cases, synthetic data may be an effective alternative. Start-ups such as **Anyverse.ai** and **Datagen.tech** focus on synthetic data development, thus making it possible to mitigate issues around data imbalance and data scarcity. However, synthetic data may be costly, and it is important that we do a cost-benefit analysis before embarking on this route.

Another problem we could face is when our collected samples are not representative enough for our model to function properly. Imagine you train your model to recognize human faces. You gather thousands of images of human faces and split your data into training and test sets. You train your model, and it predicts perfectly on your test set. However, when you ship this model as a product to the open market, you get a result such as *Figure 6.1*:

Figure 6.1 – The need for data augmentation

Surprising, right? Even though you trained the model on thousands of images, the model failed to learn to recognize faces if the axis is flipped vertically, horizontally, or otherwise. To mitigate this type of problem, we employ a technique called data augmentation. Data augmentation is a technique that we use to create new training data by altering the existing data in some way, such as randomly cropping, zooming in or out, or rotating or flipping the initial image. The underlying idea behind data augmentation is to enable our model to recognize an object in the image, even under unpredictable conditions such as the ones we saw in *Figure 6.1*.

Data augmentation is useful when we want more data samples from a limited training set; we can use data augmentation to efficiently increase the size of our dataset, hence giving our model more data to learn from. Also, because we can simulate various scenarios, our model is less likely to overfit as it learns the underlying patterns in the data rather than the noise in the data because our model learns about the data in multiple ways. Another important benefit of data augmentation is that it is a cost-saving technique that saves us from expensive and sometimes time-consuming data collection processes. We will be applying data augmentation in *Chapter 8, Handling Overfitting*, where we will be working on real-world image classification problems. Also, should you find yourself working on image or text data in the future, you may find data augmentation a very handy technique to know.

Beyond addressing issues around data imbalance and data diversity, we may also want to refine our model itself to make it suitably complex enough to identify patterns in the data, and we want to do this without overfitting the model. Here, the aim is to improve the model's quality by tweaking one or more settings, such as increasing the number of hidden layers, adding more neurons to each layer, changing the optimizer, or using a more sophisticated activation function. These settings can be tuned via experimentation until an optimal model is achieved.

We have discussed several ideas behind improving the performance of neural networks. Now, let's see how we can improve the result achieved on the Fashion MNIST dataset in *Chapter 5, Image Classification with Neural Networks*.

Fine-tuning hyperparameters of a neural network

It is important to establish a **baseline model** before making any improvements in machine learning. A baseline model is a simple model that we can use to evaluate the performance of more complex models. In *Chapter 5, Image Classification with Neural Networks*, we achieved an accuracy of 88.50% on our training data and 85.67% on our test data in just five epochs. In our quest to try to improve our model's performance, we will continue with our three-step (*build, compile,* and *fit*) process of constructing a neural network using **TensorFlow**. In each of the steps we use to build our neural network, there are settings that need to be configured before training. These settings are called **hyperparameters**. They control how the network will learn and perform, and mastering the art of fine-tuning them is an essential step in building successful deep learning models. Common hyperparameters include the number of neurons in each layer, the number of hidden layers, the learning rate, the activation functions, and the number of epochs. By iteratively experimenting with these hyperparameters, we can obtain the optimal setting best suited to our use case.

When building real-world models, especially when working on domain-specific problems, expert knowledge could prove very useful in pinpointing the best hyperparameter values for the task. Let's return to our notebook and experiment with different hyperparameters, and see whether we can beat our baseline model by tweaking one or more hyperparameters.

Increasing the number of epochs

Imagine you are trying to teach a child the multiplication tables; each study session you have with the child can be likened to an epoch in machine learning. If you only have a small number of study sessions with the child, odds are they will not fully grasp the concept of multiplication. Hence, the child will be unable to attempt basic multiplication problems. This scenario in machine learning is underfitting, where the model hasn't been able to grasp the underlying patterns in the data due to insufficient training.

On the other hand, let's say you spend a good amount of time teaching the child to memorize specific aspects of the times tables, say the 2s, 3s, and 4s. The child becomes skilled at reciting these tables; however, when faced with multiplying numbers such as 10 x 8, the child struggles. This happens because rather than understanding the principle of multiplication such that the child can apply the underlying idea when working with other numbers, the child simply memorized the examples learned during the study session. This scenario in machine learning is like the concept of overfitting, where our model performs well on the training data but fails to generalize well in new situations. In machine learning, when training our model, we need to strike a balance such that our model is trained well enough to learn the underlying patterns in our data and not to memorize the training data.

Let's see what the impact of training the model for longer will have on our result. This time, let's go for 40 epochs and observe what will happen:

```
#Step 1:  Model configuration
model=keras.Sequential([
    keras.layers.Flatten(input_shape=(28,28)),
    keras.layers.Dense(100, activation=»relu»),
    keras.layers.Dense(10,activation=»softmax»)


])


#Step 2: Compiling the model, we add the loss, optimizer and
evaluation metrics here
model.compile(optimizer='adam',
    loss=›sparse_categorical_crossentropy›,
    metrics=[‹accuracy›])
#Step 3: We fit our data to the model
history= model.fit(train_images, train_labels, epochs=40)
```

Here, we change the number of epochs in Step 3 from 5 to 40, keeping all other hyperparameters of our base model constant. The last five epochs of the output are displayed here:

```
Epoch 36/40
1875/1875 [==============================] - 5s 2ms/step - loss:
0.1356 - accuracy: 0.9493
Epoch 37/40
1875/1875 [==============================] - 5s 2ms/step - loss:
0.1334 - accuracy: 0.9503
Epoch 38/40
1875/1875 [==============================] - 4s 2ms/step - loss:
0.1305 - accuracy: 0.9502
Epoch 39/40
1875/1875 [==============================] - 4s 2ms/step - loss:
0.1296 - accuracy: 0.9512
Epoch 40/40
1875/1875 [==============================] - 4s 2ms/step - loss:
0.1284 - accuracy: 0.9524
```

Notice that, when we increase the number of epochs, it takes a much longer time to train our model. So, it can become computationally expensive when we have to train for a large number of epochs. After 40 epochs, we see that our model's training accuracy has jumped up to 0.9524, and it may appear to you that you have found a silver bullet for solving this problem. However, our goal is to ensure generalization; hence, the acid test for our model is to see how it will perform on the unseen data. Let's check out what the results look like on our test data:

```
test_loss, test_acc=model.evaluate(test_images,test_labels)
print('Test Accuracy: ', test_acc)
```

When we run the code, we arrive at an accuracy of 0.8692 on our test data. We can see that the longer the model is trained, the more accurate the model can be on the training data. However, if we train the model for too long, it will reach a point of diminishing returns, which is evident when we compare the difference in performance between the training and test set accuracy. It is critical to strike the right balance of epochs so that the model can learn and improve, but not to the point of overfitting on our training data. A practical approach could be to start with a small number of epochs, then increase the number of epochs as required. This approach can be effective; however, it can also be time-consuming as multiple experiments are required to find the optimal number of epochs.

What if we could perhaps set a rule to stop training before the point of diminishing returns? Yes, this is possible. Let's examine this idea next and see what difference it can make to our result.

Early stopping using callbacks

Early stopping is a regularization technique that can be used to prevent overfitting when training neural networks. When we hardcode the number of epochs into our model, we cannot halt training when the desired metric is attained, or when training begins to degrade or fails to improve any further. We saw this when we increased the number of epochs. However, to mitigate against this scenario, TensorFlow provides us with early stopping callbacks such that we can either use the in-built callback functions or design our own custom callbacks. We can monitor our experiments in real time with more control such that we can halt training before our model begins to overfit, when our model stops learning during training, or in line with other defined criteria. Early stopping can be invoked at various points during training. It can be applied at the start or end of training or based on attaining a specific metric.

Early stopping with built-in callbacks

Let's explore built-in callbacks for early stopping with TensorFlow:

1. We'll start by importing early stopping from TensorFlow:

    ```
    from tensorflow.keras.callbacks import EarlyStopping
    ```

2. Next, we initialize early stopping. TensorFlow allows us to pass in some arguments, which we use to create a `callbacks` object:

    ```
    callbacks = EarlyStopping(monitor='val_loss',
        patience=5, verbose=1, restore_best_weights=True)
    ```

Let's unpack some of the arguments used in our early stopping function:

- `monitor` can be used to track a metric we want to keep an eye on; in our case, we want to keep track of the validation loss. We could also switch it to track the validation accuracy. It is a good idea to track your experiment on your validation split, hence we set our `callbacks` to monitor the validation loss.

- The `patience` argument is set at 5. This means if there is no progress with regards to reducing the validation loss after five epochs, the training will end.

- We add the `restore_best_weight` argument and set it to `True`. This allows the callback to monitor the entire process and restores the weights from the best epoch found during training. If we set `restore_best_weight` to `False`, the model weights from the last training step.

- When we set `verbose` to 1, this ensures we are informed when callback actions take place. If we set `verbose` to 0, training stops but we get no output message.

There are a few other arguments that can be used here, but these ones work well enough for many instances with regard to applying early stopping.

3. We'll continue with our three-step approach in which we build, compile, and fit the model:

```
#Step 1:  Model configuration
model=keras.Sequential([
    keras.layers.Flatten(input_shape=(28,28)),
    keras.layers.Dense(100, activation=»relu»),
    keras.layers.Dense(10,activation=»softmax»)
])


#Step 2: Compiling the model, we add the loss, optimizer and
evaluation metrics here
model.compile(optimizer='adam',
    loss=›sparse_categorical_crossentropy›,
    metrics=[‹accuracy›])
#Step 3: We fit our data to the model
history= model.fit(train_images, train_labels,
    epochs=100, callbacks=[callbacks],
    validation_split=0.2)
```

`Step 1` and `Step 2` are the same steps we previously implemented. When building out the model, we trained for longer epochs. However, in `Step 3`, we made a few tweaks to accommodate our validation split and callbacks. We use 20% of our training data for validation and we pass our `callbacks` object into `model.fit()`. This ensures our early stopping callbacks interrupt the training when our validation loss stops falling. The output is as follows:

```
Epoch 14/100
1500/1500 [==============================] - 4s 2ms/step - loss:
0.2197 - accuracy: 0.9172 - val_loss: 0.3194 - val_accuracy:
0.8903
```

```
Epoch 15/100
1500/1500 [==============================] - 4s 2ms/step - loss:
0.2133 - accuracy: 0.9204 - val_loss: 0.3301 - val_accuracy:
0.8860
Epoch 16/100
1500/1500 [==============================] - 4s 2ms/step - loss:
0.2064 - accuracy: 0.9225 - val_loss: 0.3267 - val_accuracy:
0.8895
Epoch 17/100
1500/1500 [==============================] - 3s 2ms/step - loss:
0.2018 - accuracy: 0.9246 - val_loss: 0.3475 - val_accuracy:
0.8844
Epoch 18/100
1500/1500 [==============================] - 4s 2ms/step - loss:
0.1959 - accuracy: 0.9273 - val_loss: 0.3203 - val_accuracy:
0.8913
Epoch 19/100
1484/1500 [=============================>.] - ETA: 0s - loss:
0.1925 - accuracy: 0.9282 Restoring model weights from the end
of the best epoch: 14.
1500/1500 [==============================] - 4s 2ms/step - loss:
0.1928 - accuracy: 0.9281 - val_loss: 0.3347 - val_accuracy:
0.8912
Epoch 19: early stopping
```

Because we set verbose to 1, we can see that our experiment ends on epoch 19. Now, rather than worrying about how many epochs we need to train effectively, we can simply select a large number of epochs and implement early stopping. Next, we can also see that because we implemented restore_best_weights, the best weights are achieved on epoch 14 where we recorded the lowest validation loss (0.3194). With early stopping, we save compute time and take concrete steps against overfitting.

4. Let's see what our test accuracy looks like:

```
test_loss, test_acc = model.evaluate(test_images,
    test_labels)
print('Test Accuracy: ', test_acc)
```

Here, we achieved a test accuracy of 0.8847.

Now, let's see how we can write our own custom callbacks to implement early stopping.

Early stopping with custom callbacks

We can extend the capabilities of callbacks by writing our own custom callbacks for early stopping. This adds flexibility to callbacks so we can implement some desired logic during training. The TensorFlow documentation provides several ways to do this. Let's implement a simple callback to track our validation accuracy:

```
class EarlyStop(tf.keras.callbacks.Callback):
    def on_epoch_end(self, epoch, logs={}):
        if(logs.get('val_accuracy') > 0.85):
            print("\n\n85% validation accuracy has been reached.")
            self.model.stop_training = True


callback = EarlyStop()
```

For example, if we want to stop our training process when the model exceeds 85% accuracy on the validation set, we can do this by crafting our own custom callback called `EarlyStop`, which takes the `tf.keras.callbacks.Callback` parameter. We then define a function called `on_epoch_end`, which returns the logs for each epoch. We set `self.model.stop_training = True` and once the accuracy exceeds 85%, training ends and displays a message similar to what we get with `verbose` set to 1 when we used built-in callbacks. Now we can pass in our `callback` into `model.fit()` as we did with built-in callbacks. We then train our model using our three-step approach:

```
Epoch 1/100
1490/1500 [============================>.] - ETA: 0s - loss: 0.5325 -
accuracy: 0.8134/n/n 85% validation accuracy has been reached
1500/1500 [==============================] - 4s 3ms/step - loss:
0.5318 - accuracy: 0.8138 - val_loss: 0.4190 - val_accuracy: 0.8538
```

This time, at the end of the first epoch, we arrive at over 85% validation accuracy. Again, this is a smart way of achieving the desired metrics with minimal use of computational resources.

Now that we have a good grasp of how to select epochs and apply early stopping, let's now set our sights on other hyperparameters and see whether by tweaking one or more of them, we can improve our test accuracy of 88%. Perhaps we can start by trying out a more complex model.

Let's see what happens if we add more neurons to our hidden layer.

Adding neurons in the hidden layer

The hidden layers are responsible for the heavy lifting in neural networks, as we covered when discussing the anatomy of neural networks in *Chapter 5, Image Classification with Neural Networks*. Let's try out different numbers of neurons in the hidden layer. We'll define a function called `train_model` that will allow us to try out different numbers of neurons. The `train_model` function takes the `hidden_neurons` argument that represents the number of hidden neurons in the model. In addition, the function also takes training images, labels, callbacks, validation splits, and epochs. The function builds, compiles, and fits the model using these parameters:

```
def train_model(hidden_neurons, train_images, train_labels,
callbacks=None, validation_split=0.2, epochs=100):
    model = keras.Sequential([
        keras.layers.Flatten(input_shape=(28, 28)),
        keras.layers.Dense(hidden_neurons, activation=»relu»),
        keras.layers.Dense(10, activation=»softmax»)
    ])

    model.compile(optimizer=›adam›,
        loss=›sparse_categorical_crossentropy›,
        metrics=[‹accuracy›])

    history = model.fit(train_images, train_labels,
        epochs=epochs, callbacks=[callbacks] if callbacks else None,
        validation_split=validation_split)
    return model, history
```

To try out a list of neurons, we created a `for` loop to iterate over the neuron list called `neuron_values`. Then it applies the `train_model` function to build and train a model for each of the neurons in the list:

```
neuron_values = [1, 500]

for neuron in neuron_values:
    model, history = train_model(neurons, train_images,
        train_labels, callbacks=callbacks)
    print(f»Trained model with {neurons} neurons in the hidden layer»)
```

The `print` statement returns a message indicating that the model has been trained with 1 and 500 neurons respectively. Let's examine the results when we run the function, starting with the hidden layer with one neuron:

```
Epoch 36/40
1500/1500 [==============================] - 3s 2ms/step - loss:
1.2382 - accuracy: 0.4581 - val_loss: 1.2705 - val_accuracy: 0.4419
```

```
Epoch 37/40
1500/1500 [==============================] - 2s 1ms/step - loss:
1.2360 - accuracy: 0.4578 - val_loss: 1.2562 - val_accuracy: 0.4564
Epoch 38/40
1500/1500 [==============================] - 2s 1ms/step - loss:
1.2340 - accuracy: 0.4559 - val_loss: 1.2531 - val_accuracy: 0.4507
Epoch 39/40
1500/1500 [==============================] - 2s 1ms/step - loss:
1.2317 - accuracy: 0.4552 - val_loss: 1.2553 - val_accuracy: 0.4371
Epoch 40/40
1500/1500 [==============================] - 2s 1ms/step - loss:
1.2292 - accuracy: 0.4552 - val_loss: 1.2523 - val_accuracy: 0.4401
end of experiment with 1 neuron
```

From our result, the model with one neuron in the hidden layer was not complex enough to identify patterns in the data. This model performed well below 50%, which is a clear case of underfitting. Next, let's look at the result from the model with 500 neurons:

```
Epoch 11/40
1500/1500 [==============================] - 6s 4ms/step - loss:
0.2141 - accuracy: 0.9186 - val_loss: 0.3278 - val_accuracy: 0.8878
Epoch 12/40
1500/1500 [==============================] - 6s 4ms/step - loss:
0.2057 - accuracy: 0.9220 - val_loss: 0.3169 - val_accuracy: 0.8913
Epoch 13/40
1500/1500 [==============================] - 6s 4ms/step - loss:
0.1976 - accuracy: 0.9258 - val_loss: 0.3355 - val_accuracy: 0.8860
Epoch 14/40
1500/1500 [==============================] - 6s 4ms/step - loss:
0.1893 - accuracy: 0.9288 - val_loss: 0.3216 - val_accuracy: 0.8909
Epoch 15/40
1499/1500 [=============================>.] - ETA: 0s - loss: 0.1825
- accuracy: 0.9303Restoring model weights from the end of the best
epoch: 10.
1500/1500 [==============================] - 6s 4ms/step - loss:
0.1826 - accuracy: 0.9303 - val_loss: 0.3408 - val_accuracy: 0.8838
Epoch 15: early stopping
end of experiment with 500 neurons
```

We can see that the model is overfitting with more neurons. The model recorded an accuracy of 0.9303 on the training set but 0.8838 on the test set. In general, a larger hidden layer can learn more complex patterns; however, it would require more computational resources and be more prone to overfitting. When selecting the number of neurons in the hidden layer, it is important to consider the size of the training data. If we have a large training sample, we can afford to have a large number of neurons in the hidden layer. However, when the training sample is quite small, it may be better to consider working with a smaller number of neurons in the hidden layer. A larger number of neurons could lead to overfitting, as we saw in our experiment, and this architecture may perform even worse than a model with a smaller number of neurons in the hidden layer.

Another consideration to bear in mind is the type of data we are working with. When we work with linear data, a small number of hidden layers may be sufficient for our neural network. However, with non-linear data, we will need a more complex model to learn the complexities in the data. Finally, we must bear in mind that models with more neurons require longer training time. It is important to consider the trade-off between performance and generalization. As a rule of thumb, you can start training a model with a small number of neurons. This will train faster and avoid overfitting.

Alternatively, we can optimize the number of neurons in the hidden layer by identifying neurons that have little or no impact on the performance of the network. This approach is called **pruning**. This is outside the scope of the exam, so we will stop there.

Let's see the impact of adding more layers to our baseline architecture. So far, we have looked at making the model more complex and training for longer. How about we try out changing the optimizers? Let's mix things up a bit and see what happens.

Changing the optimizers

We have used the **Adam optimizer** as our default optimizer; however, there are other prominent optimizers and they all have their pros and cons. In this book and for your exam, we will focus on Adam, **stochastic gradient descent** (**SGD**), and **Root Mean Squared Propagation** (**RMSprop**). RMSprop has low memory requirements and offers an adaptive learning rate; on the flip side, it takes a much longer time to converge in comparison to Adam and SGD. RMSprop works well on training very deep networks such as **recurrent neural networks** (**RNNs**), which we will talk about later in this book.

On the other hand, SGD is another popular optimizer; it is simple to implement and efficient when the data is sparse. However, it is slow to converge and requires careful tuning of the learning rate. If the learning rate is too high, SGD will diverge; if the learning rate is too low, SGD will converge very slowly. SGD works well on a wide variety of problems and converges faster than other optimizers on large datasets, but it could sometimes converge slowly when training very large neural networks.

Adam is an improved version of SGD; it has low memory requirements, offers an adaptive learning rate, is a very efficient optimizer, and can converge to a good solution in fewer iterations than to SGD or RMSprop. Adam is also well suited for training large neural networks.

Let's try out these three optimizers and see which one works best on our dataset. We have changed the optimizer from Adam to RMSprop and SGD, and used the same architecture with built-in callbacks. We can see the results in *Figure 6.2*:

	Adam	RMSProp	SGD
Number of epochs before early stopping	13	9	39
Validation accuracy	0.8867	0.8788	0.8836
Test accuracy	0.8787	0.8749	0.8749

Figure 6.2 – The performance of different optimizers

Although Adam required more training epochs, its results were marginally better than those of the other optimizers. Of course, any of these optimizers can be used for this problem. In later chapters, we will work on real-world images that are more complex. There, we will revisit these optimizers.

Before we close this chapter, let's look at the learning rate and its impact on our model's performance.

Changing the learning rate

Learning rate is an important hyperparameter that controls how well our model will learn and improve during training. An optimal learning rate will ensure the model converges quickly and accurately, while on the other hand, a poorly selected learning rate can lead to a wide range of issues such as slow convergence, underfitting, overfitting, or network instability.

To understand the impact of the learning rate, we need to know how it affects the model's training process. The learning rate is the step size taken to reach the point where the loss function is at its minimum. In *Figure 6.3(a)*, we see that when we choose a low learning rate, the model requires too many steps to reach the minimum point. On the flip side, when the learning rate is too high, the model would likely learn too quickly, taking larger steps and likely overshooting the minimum point, as seen in *Figure 6.3(c)*. A high learning rate can lead to instability and overfitting. However, when we find the ideal learning rate as in *Figure 6.3(b)*, the model is likely to experience fast convergence and good generalization:

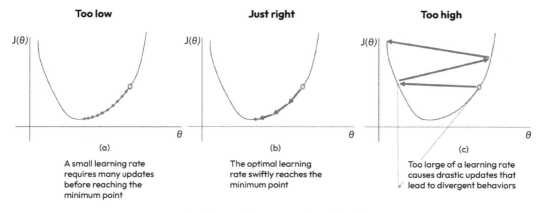

Figure 6.3 – A plot showing low, optimal, and high learning rates

The question that comes to mind is: how do we find the optimal learning rate? One way is to try out different learning rates and see what works based on evaluating the model on the validation set. Another way is to use a learning rate scheduler. This allows us to dynamically adjust the learning rate during training. We will explore this approach in the later chapters of the book. Here, let's try out a few different learning rates to see how they impact on our network.

Let's craft a function that will take a list of different learning rates. In this experiment, we will try out six different learning rates (1, 0.1, 0.01, 0.001, 0.0001, 0.00001, and 0.000001). First, let's create a function to create our model:

```
def learning_rate_test(learning_rate):
    #Step 1:  Model configuration
    model=keras.Sequential([
        keras.layers.Flatten(input_shape=(28,28)),
        keras.layers.Dense(64, activation=»relu»),
        keras.layers.Dense(10,activation=»softmax»)


])
    #Step 2: Compiling the model, we add the loss,
        #optimizer and evaluation metrics here
    model.compile(optimizer=tf.keras.optimizers.Adam(
        learning_rate=learning_rate),
        loss='sparse_categorical_crossentropy',
        metrics=[<accuracy>])
    #Step 3: We fit our data to the model
    callbacks = EarlyStopping(monitor='val_loss',
        patience=5, verbose=1, restore_best_weights=True)
    history=model.fit(train_images, train_labels,
        epochs=50, validation_split=0.2,
        callbacks=[callbacks])
    score=model.evaluate(test_images, test_labels)
    return score[1]
```

We will use the function to build, compile, and fit the model. It also takes in the learning rate as a variable that we pass into our function, and that returns the test accuracy as our result:

```
# Try out different learning rates
learning_rates = [1, 0.1, 0.01, 0.001, 0.0001, 0.00001,
    0.000001]

# Create an empty list to store the accuracies
accuracies = []

# Loop through the different learning rates
```

```
for learning_rate in learning_rates:
    # Get the accuracy for the current learning rate
    accuracy = learning_rate_test(learning_rate)

    # Append the accuracy to the list
    accuracies.append(accuracy)
```

We have now outlined the different learning rates. Here, we want to experiment with different learning rates, from a very high to a very low learning rate. We created an empty list and appended our test set accuracy. Next, let's look at the numeric values in a tabular fashion. We use pandas to generate a DataFrame with the learning rate and accuracies:

```
df = pd.DataFrame(list(zip(learning_rates, accuracies)),
    columns =[<Learning_rates>, <Test_Accuracy>])
df
```

The output DataFrame is shown in the following screenshot:

	Learning_rates	Test_Accuracy
0	1.000000	0.1000
1	0.100000	0.3487
2	0.010000	0.8567
3	0.001000	0.8809
4	0.000100	0.8748
5	0.000010	0.8429
6	0.000001	0.7731

Figure 6.4 – Different learning rates and their test accuracies

From the results, we can see that when working with a really high learning rate (1.0), the model performs poorly. As we reduce the learning rate value, we see that the model's accuracy begins to rise; when the learning rate becomes too small, the model takes too long to converge. There is no silver bullet when it comes to choosing the ideal learning rate for a problem. It depends on several factors such as the model architecture, the data, and the type of optimization technique applied.

Now that we have seen various ways of tweaking a model to improve its performance, we have come to the end of the chapter. We've tried adjusting different hyperparameters to improve our model's performance; however, we got stuck on 88% test accuracy. Perhaps this is a good time to try something else, which we will do in the next chapter. Take a break, and when you are ready, let's see how we can improve this result and also try out real-world images.

Summary

In this chapter, we looked at improving the performance of a neural network. Although we worked with a lightweight dataset, we have learned some important ideas around improving our model's performance–ideas that will come in handy, both in the exam and on the job. You now know that data quality and model complexity are two sides of the machine learning coin. If you have good-quality data, a poor model will yield subpar results and, on the flip side, even the most advanced model will yield a suboptimal result with bad data.

By now, you should have a good understanding and hands-on experience of fine-tuning neural networks. Like a seasoned expert, you should be able to understand the art of fine-tuning hyperparameters and apply this to different machine learning problems and not just image classification. Also, you have seen that model building requires a lot of experimenting. There is no silver bullet, but having a good understanding of the moving parts and various techniques, and how and why to apply them, is what differentiates a star from the average Joe.

In the next chapter, we will examine convolutional neural networks. We will see why they are state-of-the-art when it comes to image classification tasks. We will look at the power of convolutions and examine in a hands-on fashion how they do things differently from the simple neural networks we have been using so far.

Questions

Let's test what we've learned in this chapter using the CIFAR-10 notebook:

1.   Build a neural network using our three-step approach.

2.   Increase the number of neurons from 5 to 100 in the hidden layer.

3.   Use a custom callback to stop training when the training accuracy is 90%.

4.   Try out the following learning rates: 5, 0.5, 0.01, 0.001. What did you observe?

Further reading

To learn more, you can check out the following resources:

- Amr, T., 2020. *Hands-On Machine Learning with scikit-learn and Scientific Python Toolkits*, Packt Publishing.

- Gulli, A., Kapoor, A. and Pal, S., 2019. *Deep Learning with TensorFlow 2 and Keras*, Packt Publishing.

- *How to Write Custom TensorFlow Callbacks — The Easy Way*: `https://towardsdatascience.com/how-to-write-custom-tensorflow-callbacks-the-easy-way-c7c4b0e31c1c`

- https://medium.com/geekculture/introduction-to-neural-network-2f8b8221fbd3

- https://www.tensorflow.org/api_docs/python/tf/keras/callbacks/EarlyStopping

7

Image Classification with Convolutional Neural Networks

Convolutional neural networks (**CNNs**) are the go-to algorithms when it comes to image classification. In the 1960s, neuroscientists Hubel and Wiesel conducted a study on the visual cortex in cats and monkeys. Their work unraveled how we visually process information in a hierarchical structure, showing how visual systems are organized into a series of layers where each layer is responsible for a different aspect of visual processing. This earned them a Nobel Prize, but more importantly, it served as the basis upon which CNNs are built. CNNs, by virtue of their nature, are well designed to work with data with spatial structures such as images.

However, in the early days, CNNs did not have the limelight due to a number of factors, such as insufficient training data, underdeveloped network architecture, insufficient computational resources, and the absence of modern techniques such as data augmentation and dropout. In the 2012 ImageNet Large Scale Visual Recognition Challenge, the **machine learning** (**ML**) community was taken by storm when a CNN architecture called AlexNet outperformed all other methods by a large margin. Today, ML practitioners apply CNNs to achieve state-of-the-art performance on computer vision tasks such as image classification, image segmentation, and object detection, among others.

In this chapter, we will examine CNNs to see how they do things differently from the fully connected neural networks we have used so far. We will start with the challenges faced by fully connected networks when working with image data, after which we will explore the anatomy of CNNs. We will look at the core building blocks of CNN architecture and their overall impact on the performance of the network. Next, we will build an image classifier using a CNN architecture with the Fashion MNIST dataset, then move on to building a real-world image classifier. We will be working with color images of different sizes, and our target objects are in different positions within the image.

By the end of this chapter, you will have a sound understanding of what CNNs are and why they are superior to fully connected networks when it comes to image classification tasks. Also, you will be able to effectively build, train, tune, and test CNN models on real-world image classification problems.

In this chapter, we will cover the following topics:

- The anatomy of CNNs

- Fashion MNIST with CNNs

- Real-world images

- Weather data classification

- Applying hyperparameters to improve the model's performance

- Evaluating image classifiers

Challenges of image recognition with fully connected networks

In *Chapter 5, Image Classification with Neural Networks*, we applied a **deep neural network (DNN)** to the Fashion MNIST dataset. We saw how every neuron in the input layer is connected to every neuron in the hidden layer and those in the hidden layer are connected to neurons in the output layer, hence the name *fully connected*. While this architecture can solve many ML problems, they are not well suited for modeling image classification tasks, due to the spatial nature of image data. Let's say you are looking at a picture of a face; the positioning and orientation of the features on the face enable you to know it is a human face even when you just focus on a specific feature, such as the eyes. Instinctively, you know it's a face by virtue of the spatial relationship between the features of the face; however, DNNs do not see this bigger picture when looking at images. They process each pixel in the image as independent features, without taking the spatial relationships between these features into consideration.

Another issue with using fully connected architectures is the curse of dimensionality. Let's say we are working with a real-world image of size 150 x 150 with 3 color channels, **red, green, and blue (RGB)**; we will have an input size of 67,500. As all the neurons are connected to neurons in the next layer, if we feed these values into a hidden layer with 500 neurons, we will have 67,500 x 500 = 33,750,000 parameters, and this number of parameters will grow exponentially as we add more layers, making it resource intensive to apply this type of network to image classification tasks. Another accompanying problem we could stumble upon is overfitting; this happens due to the large number of parameters in our network. If we have images of larger sizes or we add more neurons to our networks, the number of trainable parameters will grow exponentially, and it could become impractical to train such a network due to cost and resource requirements. In light of these challenges, there is a need for a more sophisticated architecture, and this is where CNNs come in with their ability to uncover spatial relationships and hierarchies, ensuring features are recognized irrespective of where they are located within an image.

Note

Spatial relationship refers to how features within an image are arranged in relation to each other in terms of position, distance, and orientation.

Anatomy of CNNs

In the last section, we saw some of the challenges DNNs grappled with when dealing with visual recognition tasks. These issues include the lack of spatial awareness, high dimensionality, computational inefficiency, and the risk of overfitting. How do we overcome these challenges? This is where CNNs come into the picture. CNNs by design are uniquely positioned to handle image data. Let's go through *Figure 7.1* and uncover why and how CNNs stand out:

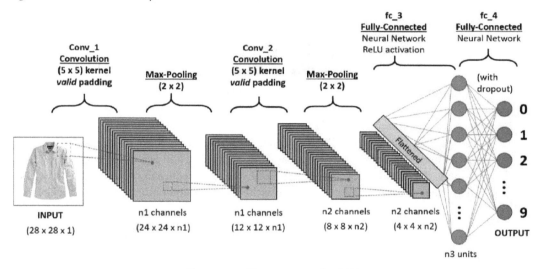

Figure 7.1 – The anatomy of a CNN

Let's break down the different layers in the diagram:

1. **Convolutional layer – the eyes of the network**: Our journey begins with us feeding in images into the convolutional layer; this layer can be viewed as the "eyes of our network." Their job is primarily to extract vital features. Unlike DNNs, where each neuron is connected to every neuron in the next layer, CNNs apply filters (also known as kernels) to capture local patterns within an image in a hierarchical fashion. The output of the interactions between a segment of the input image that the filter slides over is called a feature map. As shown in *Figure 7.2*, we can see that each feature map highlights specific patterns in the shirt that we passed into the network. Images go through CNNs in a hierarchical fashion with filters in the earlier layers adept at capturing simple features, while those in subsequent layers capture more complex patterns, mimicking the hierarchical structure of a human's visual cortex. Another important property of CNNs is

parameter sharing – this happens because patterns are only learned once and applied everywhere else across an image. This ensures that the visual ability of the model is not location-specific. In ML, we refer to this concept as **translation invariance** – the network's ability to detect a shirt regardless of whether it is aligned to the right or left or centered within an image.

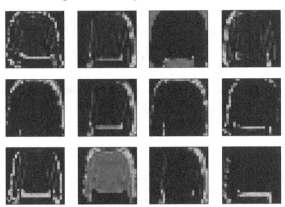

Figure 7.2 – Visualization of features captured by a convolutional layer

2. **Pooling layer – the summarizer**: After the convolutional layer comes the pooling layer. This layer can be viewed as a summarizer in CNNs as it focuses on condensing the overall dimensionality of the feature maps while retaining important features, as illustrated in *Figure 7.3*. By methodically downsampling the feature maps, CNNs significantly not only reduce the number of parameters required for image processing but also improve the overall computational efficiency of CNNs.

Original Image

Pooling Output

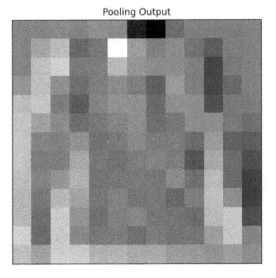

Figure 7.3 – An example of the pooling operation, preserving essential details

3. **Fully connected layer – the decision maker**: Our image traverses a series of convolution and pooling layers that extract features and reduce the dimensionality of feature maps, eventually reaching the fully connected layer. This layer can be viewed as the decision maker. This layer offers high-level reasoning as it brings together all the important details collected through the layers and uses them to make the final classification verdict. One of the hallmarks of CNNs is its end-to-end learning process, which seamlessly integrates feature extraction and image classification. This methodological and hierarchical learning approach makes CNN a well-suited tool for image recognition and analysis.

We have only scratched the surface of how CNNs work. Let's now drill down into the key operations that take place within the different layers, starting with convolutions.

Convolutions

We now know that convolutional layers apply filters, which slide over patches of the input image. A typical CNN applies multiple filters, with each filter learning a specific kind of feature by interacting with the input image. By combining the detected features, a CNN arrives at a comprehensive understanding of the image features and uses this detailed information to classify the input image. Mathematically, this convolution process involves the dot product between a patch of the input image and the filter (a small matrix), as illustrated in *Figure 7.4*. This process yields an output known as the **activation map** or **feature map**.

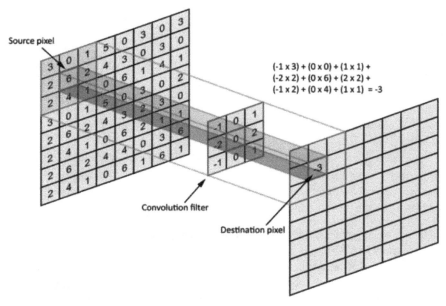

The convolution operation.

Figure 7.4 – Convolution operation – applying a filter to an input image to generate a feature map

As the filter slides over the patches of the image, it produces a feature map for each dot operation. Feature maps are a representation of the input image in which certain visual patterns are enhanced by the filter, as shown in *Figure 7.2*. When we stack feature maps from all the filters in the network, we arrive at a rich, multi-faceted view of the input image, which gives later layers adequate information to learn more complex patterns.

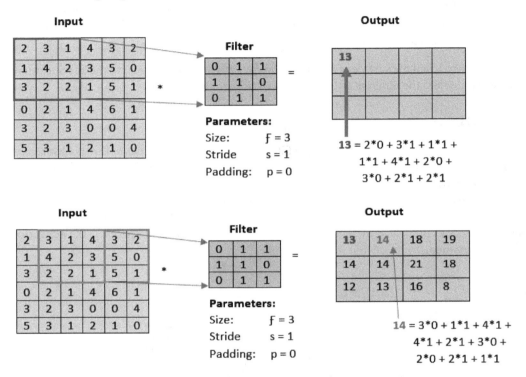

Figure 7.5 – a (top) and b (bottom): Dot product computation

In *Figure 7.5 a*, we see a dot product operation in progress, as a filter slides over a section of the input image resulting in a destination pixel value of 13. If we move the filter 1 pixel to the right, as shown in *Figure 7.5 b*, we will arrive at the next destination pixel value of 14. If we continue sliding the filter one pixel at a time over the input image, we will achieve the complete output shown in *Figure 7.5 b*.

We have now seen how convolution operations work; however, there are various types of convolutional layers that we can apply in CNNs. For image classification, we typically use 2D convolutional layers, while we apply 1D convolution layers for audio processing and 3D convolutional layers for video processing. When designing our convolutional layer, there are a number of adjustable hyperparameters that can impact the performance of our network, such as the number of filters, the size of the filters, stride, and padding. It is pertinent to explore how these hyperparameters impact our network.

Let's begin this exploration by looking at the impact of the number of filters in a convolutional layer.

Impact of the number of filters

By increasing the number of filters within a CNN, we empower it to learn a richer and more diverse representation of the input image. The more filters we have, the more representation will be learned. However, more filters mean more parameters to train, and this could not only increase the computational cost but also slow down the training process and increase the risk of overfitting. When deciding on the number of filters to apply to your network, it is important to consider the type of data in use. If the data has a lot of variability, you may need more filters to capture the diversity in your data, whereas with smaller datasets, you should be more conservative to reduce the risk of overfitting.

Impact of the size of the filter

We now know that filters are small matrices that slide over our input image to produce feature maps. The size of the filter we apply to the input image will determine the level and type of features that will be extracted from the input image. The filter size is the dimension of the filter – that is, the height and width of the filter matrix. Typically, you will come across 3x3, 5x5, and 7x7 filters. Smaller filters will cover a smaller patch of the input image, while a larger filter will cover a more extensive section of the input image:

- **Granularity of features** – Smaller filters such as 3x3 filters can be applied to capture finer and more local details of an image such as edges, textures, and corners, while larger filters such as 7x7 filters can learn broader patterns such as face shapes or object parts.

- **Computational efficiency** – Smaller filters cover a smaller receptive field of the input image, as illustrated in *Figure 7.6*, which means they will require more operations.

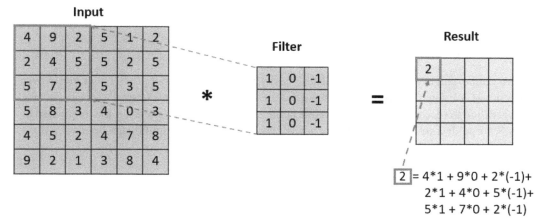

Parameters: Size: f = 3 | Stride: s = 1 | Padding: p = 0

Figure 7.6 – Convolution operation with a 3x3 filter

On the other hand, a larger filter covers a large segment of the input image, as shown in *Figure 7.7*. However, many modern CNN architectures (for example, VGG) use 3x3 filters. Stacking these smaller filters together would increase the depth of the network and enhance the capabilities of these filters to capture more complex patterns with a smaller number of parameters in comparison to using a large filter, which makes smaller filters easier to train.

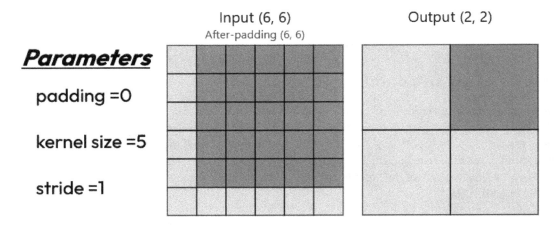

Figure 7.7 – Convolution operation with a 5x5 filter

- **Parameter count** – A larger filter typically has more weight in comparison to a smaller filter; for example, a 5x5 filter will have 25 parameters and a 3x3 filter will have 9 parameters. Here, we are ignoring the depth for the sake of simplicity. Hence, larger filters will contribute to making the model more complex in comparison to smaller filters.

Impact of stride

Stride is an important hyperparameter in CNNs. It determines the number of pixels a filter moves over an input image. We can liken stride to the step we take when walking; if we take small steps, it will take us a longer time to reach our destination, while larger steps will ensure we reach it much quicker. In *Figure 7.8*, we apply a stride of 1, which means the filter moves over the input image 1 pixel at a time.

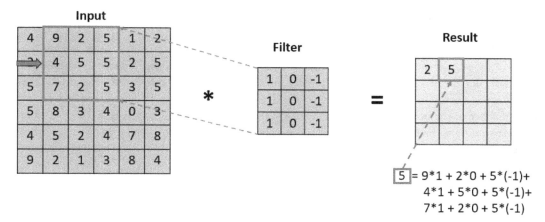

Parameters: Size: f = 3 | Stride: s = 1 | Padding: p = 0

Figure 7.8 – Convolution operation with a stride of 1

If we apply a stride of 2, it means the filter will move 2 pixels at a time, as illustrated in *Figure 7.9*. We see that a large stride will lead to a reduced spatial dimension of the output feature map. We can see this when we compare the output of both figures.

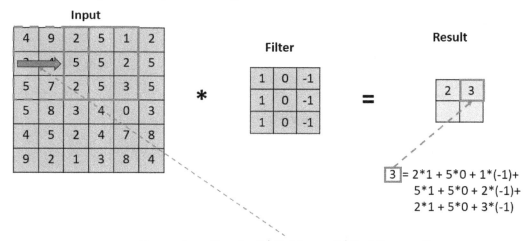

Parameters: Size: f = 3 | Stride: s = 2 | Padding: p = 0

Figure 7.9 – Convolution operation with a stride of 2

When we apply a larger stride, it can increase the computational efficiency, but it also reduces the spatial resolution of the input image. Hence, we need to consider this trade-off when selecting the right stride for our network. Next, let's examine the border effect.

The boundary problem

When a filter slides over the input image performing convolution operations, it soon reaches the borders or the edges, where it becomes difficult to perform dot product operations due to the absence of pixels outside the image boundaries. This results in the output feature map being smaller than the input image as a result of loss of information around the edges or borders. This issue is referred to as the **edge effect** or the **boundary problem** in ML. In *Figure 7.10*, we can observe that we are unable to perform a dot product operation on the bottom-left corner as we cannot center the filter over the highlighted pixel value of 3 without some part of the filter falling out of the defined image boundary.

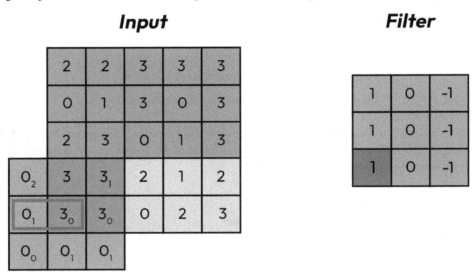

Figure 7.10 – Showing the boundary problem

To fix the boundary issue and preserve the spatial dimension of the output feature map, we may want to apply padding. Let's discuss this concept next.

Impact of padding

Padding is a technique we can apply to our convolution process to prevent the boundary effect by adding extra pixels to the edges, as shown in *Figure 7.11*.

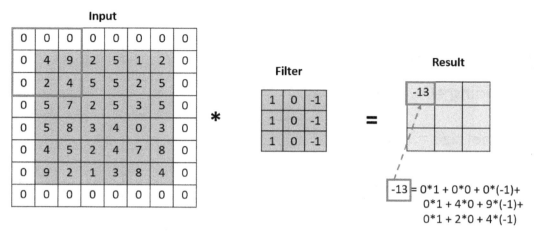

Parameters: Size: f = 3 | Stride: s = 1 | Padding: p = 0

Figure 7.11 – A padded image undergoing convolution operation

We can now perform dot product operations on pixels at the edges, hence preserving information at the edges. Padding can also be applied to maintain the spatial dimension pre- and post-convolution. This could prove useful in deep CNN architecture with several convolution layers. Let's look at the two main types of padding:

- **Valid Padding (No Padding)**: Here, no padding is applied. This can be useful when we want to achieve a reduced spatial dimensionality, especially in deeper layers.

- **Same Padding**: Here, we set padding to ensure the output feature map and the input image dimensions are the same. We use this when maintaining spatial dimensionality is of paramount importance.

Before we move on to examining the pooling layer, let's put together the different hyperparameters we have discussed in the convolutional layer and see them in action.

Putting it all together

In *Figure 7.12*, we have a 7x7 input image and a 3x3 filter. Here, we use a stride of 1 and set padding to Valid (no padding).

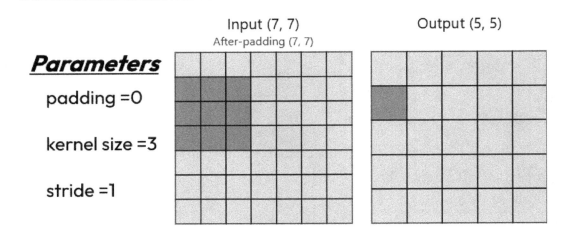

Figure 7.12 – Setting the hyperparameters

To compute the output feature map of a convolutional operation, we can apply the following formula:

$$\left(\frac{W - F + 2P}{S}\right) + 1$$

In this formula, the following applies:

- W represents the size of the input image

- F stands for the filter size

- S represents stride

- P stands for padding

When we input the respective values into the equation, we get a resulting value of 5, which means we will have a 5x5 output feature map. If we alter any of the values, it will impact the size of the output feature map one way or the other. For example, if we increase the stride size, we will have a smaller output feature map, while if we set padding to same, this will increase the size of the output. We can now move on from the convolution operations and explore pooling next.

Pooling

Pooling is an important operation that takes play in the pooling layer of a CNN. It is a technique used to downsample the spatial dimension of individual feature maps generated by the convolutional layers. Let's examine some important types of pooling layers. We'll begin by exploring max pooling, as shown in *Figure 7.13*. Here, we see how max pooling operations work. The pooling layer simply takes the highest value from each region of the input data.

Max Pooling

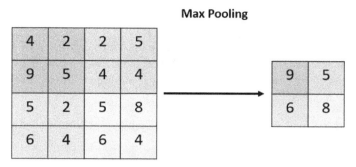

Figure 7.13 – A max pooling operation

Max pooling enjoys several benefits as it is intuitive and easy to implement. It is also efficient since it simply extracts the highest value in a region, and it has been applied with good effect across diverse tasks.

Average pooling, as the name suggests, reduces the data dimensionality by taking the average value for a designated region, as illustrated in *Figure 7.14*.

Average Pooling

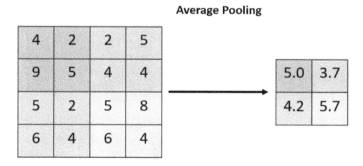

Figure 7.14 – An average pooling operation

On the other hand, min pooling extracts the minimum value in a specified region of the input data. Pooling reduces the spatial size of the output feature maps and this, in turn, reduces the memory requirement for storing intermediate representations. Pooling can be beneficial to a network; however, excessive pooling can be counterproductive as this could lead to information loss. After the pooling layer, we arrive at the fully connected layer, the decision maker of our network.

The fully connected layer

The final component of our CNN architecture is the fully connected layer. Unlike the convolutional layer, here, every neuron is connected to every neuron in the next layer. This layer is responsible for decision-making, such as classifying whether our input image is a shirt or a hat. The fully connected layer takes the learned features from the earlier layers and maps them to their corresponding labels. We have now covered CNNs in theory; let's now proceed by applying them to our fashion dataset.

Fashion MNIST 2.0

By now, you are already familiar with this dataset, as we used it in *Chapter 5, Image Classification with Neural Networks*, and *Chapter 6, Improving the Model*. Now, let's see how CNNs compare to the simple neural networks we have worked with so far. We will continue in the same spirit as before. We start by importing the required libraries:

1. We will import the requisite libraries for preprocessing, modeling, and visualizing our ML model using TensorFlow:

    ```
    import tensorflow as tf
    import numpy as np
    import matplotlib.pyplot as plt
    ```

2. Next, we will load the Fashion MNIST dataset from TensorFlow Datasets using the `load_data()` function. This function returns our training and testing data consisting of NumPy arrays. The training data consists of `x_train` and `y_train`, and the test data is made up of `x_test` and `y_test`:

    ```
    (x_train,y_train),(x_test,y_test) = tf.keras.datasets.fashion_
    mnist.load_data()
    ```

3. We can confirm the data size by using the `len` function on our training and testing data:

    ```
    len(x_train), len(x_test)
    ```

    When we run the code, we get the following output:

    ```
    (60000, 10000)
    ```

    We can see that we have a training data size of 60,000 images and test data of 10,000 images.

4. In CNNs, unlike the DNNs we used previously, we need to account for the color channels of the input images. Currently, our training and testing data has a shape of (`batch_size`, `height`, `width`) for grayscale images, with a single channel. However, CNN models require a 4D input tensor, made up of `batch_size`, `height`, `width`, and `channels`. We can fix this data mismatch by simply reshaping our data and converting the elements to `float32` values:

    ```
    # Reshape the images(batch_size, height, width, channels)
    x_train = x_train.reshape(x_train.shape[0],
        28, 28, 1).astype('float32')
    x_test = x_test.reshape(x_test.shape[0],
        28, 28, 1).astype('float32')
    ```

    This preprocessing step is standard before training an ML model, as most models require floating-point input. Since our images are grayscale, there is only one color channel, which is why we reshape the data to include a single channel dimension.

5. The pixel value of our data (training and testing data) ranges from 0 to 255, where 0 represents black and 255 represents white. We normalize our data by dividing the pixel values by 255 to bring the pixel values in our data to a scale of between 0 and 1. We do this to enable our model to converge faster and perform better:

```
# Normalize the pixel values
x_train /= 255
x_test /= 255
```

6. We use the `to_categorical` function from the `utils` module of `tf.keras` to convert our labels (`y_train` and `y_test`) that have an integer value of 0 to 9 into one-hot encoded arrays. The `to_categorical` function takes two arguments: the labels to be converted, and the number of classes; it returns a one-hot encoded array, as shown in *Figure 7.15*.

Before
y_train [0] = 9

After
y_train [0] = [0.,0.,0.,0.,0.,0.,0.,0.,0.,1.]

Figure 7.15 – A one-hot encoded array

The one-hot encoded vectors will have a length of 10, with a number 1 in the index that corresponds to the label for a given data point, and 0 in all other indices:

```
# Convert the labels to one hot encoding format
y_train = tf.keras.utils.to_categorical(y_train, 10)
y_test = tf.keras.utils.to_categorical(y_test, 10)
```

7. Using the Sequential Model API from `tf.keras.model`, we will create a CNN architecture:

```
# Build the Sequential model
model = tf.keras.models.Sequential()
# Add convolutional layer
model.add(tf.keras.layers.Conv2D(64,kernel_size=(3,3),
    activation='relu',
    input_shape=(28, 28, 1)))
# Add max pooling layer
model.add(tf.keras.layers.MaxPooling2D(pool_size=(2, 2)))

# Flatten the data
model.add(tf.keras.layers.Flatten())

# Add fully connected layer
model.add(tf.keras.layers.Dense(128,
                                activation='relu'))
```

```
# Apply softmax
model.add(tf.keras.layers.Dense(10,
                                activation='softmax'))
```

The first layer is a convolution layer composed of 64 filters of size 3x3 to process the input images, which have a shape of 28x28 pixels and 1 channel (grayscale). ReLU is used as the activation function. The subsequent max pooling layer is a 2D pooling layer that applies max pooling to downsample the output of the convolution layer, reducing the dimensionality of the feature maps. The flatten layer takes the output of the pooling layer and flattens it into a 1D array, which is then processed by the fully connected layer. The output layer contains softmax activation for multiclass classification and 10 neurons, one for each class.

8. Next, we compile and fit the model on our training data:

```
# Compile and fit the model
model.compile(loss='categorical_crossentropy',
              optimizer='adam', metrics=['accuracy'])
model.fit(x_train, y_train, epochs=10,
          validation_split=0.2)
```

The compile() function takes three arguments: loss function (categorical_crossentropy, since this is a multi-class classification task), optimizer (adam), and metrics (accuracy). After compiling the model, we used the fit() function to train the model on the training data. We specified the number of epochs as 10 and used 20% of the training data for validation purposes.

In 10 epochs, we arrive at a training accuracy of 0.9785 and a validation accuracy of 0.9133:

```
Epoch 6/10
1500/1500 [==============================] - 5s 3ms/step - loss:
0.1267 - accuracy: 0.9532 - val_loss: 0.2548 - val_accuracy:
0.9158
Epoch 7/10
1500/1500 [==============================] - 5s 4ms/step - loss:
0.1061 - accuracy: 0.9606 - val_loss: 0.2767 - val_accuracy:
0.9159
Epoch 8/10
1500/1500 [==============================] - 6s 4ms/step - loss:
0.0880 - accuracy: 0.9681 - val_loss: 0.2957 - val_accuracy:
0.9146
Epoch 9/10
1500/1500 [==============================] - 6s 4ms/step - loss:
0.0697 - accuracy: 0.9749 - val_loss: 0.3177 - val_accuracy:
0.9135
Epoch 10/10
1500/1500 [==============================] - 6s 4ms/step - loss:
0.0588 - accuracy: 0.9785 - val_loss: 0.3472 - val_accuracy:
0.9133
```

9. The `summary` function is a very useful way to get a high-level overview of the model's architecture and understand the number of parameters and the shape of the output tensors:

```
model.summary()
```

The output returns the five layers that make up our current model architecture. It also displays the output shape and the number of parameters of each layer. The total number of parameters is 1,386,506. From the output, we see that the output shape from the convolution layer is 26x26 as a result of the border effect since we did not apply padding. Next, the max pooling layer halves the pixel size after which we flatten the data and generate predictions:

```
Model: "sequential"

_____
 Layer (type)                Output Shape              Param #
=================================================================
 conv2d (Conv2D)             (None, 26, 26,64)         640

 max_pooling2d (MaxPooling2D  (None,13,13,64)          0        )

 flatten (Flatten)           (None, 10816)             0

 dense (Dense)               (None, 128)               1384576

 dense_1 (Dense)             (None, 10)                1290


=================================================================
Total params: 1,386,506
Trainable params: 1,386,506
Non-trainable params: 0
_____
```

10. Finally, we will use the `evaluate` function to evaluate our model on test data. The `evaluate` function returns the loss and the accuracy of the model on test data:

```
# Evaluate the model
score = model.evaluate(x_test, y_test)
```

Our model achieved an accuracy of 0.9079 on the test data, surpassing the performance of the architectures used in *Chapter 6, Improving the Model*. We can try to further improve the model's performance by adjusting the hyperparameters and applying data augmentation. Let's turn our attention to real-world images, where CNNs clearly outshine our previous models.

Working with real-world images

Real-world images pose a different type of challenge as these images are usually colored images with three color channels (red, green, and blue), unlike the grayscale images we used from our fashion MNIST dataset. In *Figure 7.16*, where we see an example of real-world images from the weather dataset that we will be modeling shortly, you will notice the images are of varying sizes. This introduces another layer of complexity that requires additional preprocessing steps such as resizing or cropping to ensure all our images are of uniform dimensions before we feed them into our neural network.

Figure 7.16 – Images from the weather dataset

Another issue we may encounter when working with real-world images is the presence of various noise sources. For example, we may have images in our dataset taken in conditions with uneven lighting or unintended blurring. Again, we could have images with multiple objects or other unintended distractions in the background among the images in our real-world dataset.

To address these issues, we could apply noise reduction techniques such as denoising to improve the quality of our data. We could also use object detection techniques such as bounding boxes or segmentation to help us identify the target object within an image with multiple objects. The good part is TensorFlow is well equipped with a comprehensive set of tools tailored to handling these challenges. One important tool from TensorFlow is the `tf.image` module, which offers an array of image preprocessing functionalities such as resizing various adjustments (for example, brightness, contrast, hue, and saturation), application of bounding boxes, cropping, flipping, and much more.

However, this module is beyond the scope of this book and the exam itself. But, if you wish to learn more about this module, you can visit the TensorFlow documentation at `https://www.tensorflow.org/api_docs/python/tf/image`. Another tool in TensorFlow's arsenal is `ImageDataGenerator`, which enables us to perform data augmentation on the fly, offering us the ability to preprocess and perform augmentative actions (such as rotation and flipping images) in real time as we feed these images into our training pipeline. Let's proceed to work with our real-world image dataset and see `ImageDataGenerator` in action.

Weather dataset classification

In this case study, we will be working as a computer vision consultant for an emerging start-up called WeatherBIG. You have been assigned the responsibility of developing an image classification system that will be used to identify different weather conditions; the dataset for this task can be found on Kaggle using this link: https://www.kaggle.com/datasets/rahul29g/weatherdataset. The dataset has been packaged into three folders made up of a training folder, a validation folder, and a testing folder. Each of these folders has subfolders with each weather class. Let's get started with the task:

1. We start by importing several libraries to build our image classifier:

```
import os
import pathlib
import matplotlib.pyplot as plt
import matplotlib.image as mpimg
import random
import numpy as np
from PIL import Image
import tensorflow as tf
from tensorflow import keras
from tensorflow.keras.preprocessing.image import
ImageDataGenerator
```

We have used several of these libraries in our previous experiments; however, let's address the functionalities of a few libraries that we will be using for the first time. The `os` module acts as a bridge to our operating system. It gives us the ability to read from and write to our filesystem, while `pathlib` offers us an intuitive, object-oriented way to streamline our file navigation tasks. For image manipulation, we use `PIL`, and we also have the `ImageDataGenerator` class from the `tensorflow.keras.preprocessing.image` module for our data preprocessing steps, batch generation, and data augmentation.

2. You can access/download the dataset for this case study from https://www.kaggle.com/datasets/rahul29g/weatherdataset and upload it to Google Drive. Once you do this, you can easily follow along with the code in this section. In my case, the data is stored in this root directory: `/content/drive/MyDrive/weather dataset`. In your case, your root directory will be different, so make sure you change the directory path to match the directory where the dataset is stored in your Google Drive: `root_dir = "/content/drive/MyDrive/weather dataset"`.

3. Next, we apply the `os.walk` function to access the root directory and generate information about the content of all the directories and subdirectories:

```
for dirpath, dirnames, filenames in os.walk(root_dir):
    print(f"Directory: {dirpath}")
    print(f"Number of images: {len(filenames)}")
    print()
```

Running the code returns a tuple made up of the path of each directory and the number of images within each of them, as illustrated in *Figure 7.17*:

```
Directory: /content/drive/MyDrive/weather dataset
Number of images: 0

Directory: /content/drive/MyDrive/weather dataset/train
Number of images: 0

Directory: /content/drive/MyDrive/weather dataset/train/cloud
Number of images: 207

Directory: /content/drive/MyDrive/weather dataset/train/shine
Number of images: 189

Directory: /content/drive/MyDrive/weather dataset/train/rain
Number of images: 152

Directory: /content/drive/MyDrive/weather dataset/train/sunrise
Number of images: 243
```

Figure 7.17 – A snapshot directory and subdirectories

We use this step to get a sense of the contents of each directory and subdirectory.

4. We use the `retrieve_labels` function to fetch and display labels and their corresponding counts from the training, test, and validation directories. To craft this function, we use the `listdir` method from the `os` module and we pass in the respective directory paths (`train_dir`, `test_dir`, and `val_dir`):

```
def retrieve_labels(train_dir, test_dir, val_dir):
    # Retrieve labels from training directory
    train_labels = os.listdir(train_dir)
    print(f"Training labels: {train_labels}")
    print(f"Number of training labels: {len(train_labels)}")
    print()

    # Retrieve labels from test directory
    test_labels = os.listdir(test_dir)
    print(f"Test labels: {test_labels}")
    print(f"Number of test labels: {len(test_labels)}")
    print()

    # Retrieve labels from validation directory
    val_labels = os.listdir(val_dir)
    print(f"Validation labels: {val_labels}")
    print(f"Number of validation labels: {len(val_labels)}")
    print()
```

5. We specify the path to the training, test, and validation directories in the `train_dir`, `test_dir`, and `val_dir` arguments, respectively:

```
train_dir = "/content/drive/MyDrive/weather dataset/train"
test_dir = "/content/drive/MyDrive/weather dataset/test"
val_dir = "/content/drive/MyDrive/weather dataset/validation"
retrieve_labels(train_dir, test_dir, val_dir)
```

When we run the code, it returns the training data, test data, validation data labels, and the number of labels:

```
Training labels: ['cloud', 'shine', 'rain', 'sunrise']
Number of training labels: 4

Test labels: ['sunrise', 'shine', 'cloud', 'rain']
Number of test labels: 4

Validation labels: ['shine', 'sunrise', 'cloud', 'rain']
Number of validation labels: 4
```

6. For our exploration, let's craft a function called `view_random_images` to randomly access and display images from the subdirectories within our dataset. The function takes in the main directory that holds the subdirectories housing our images and the number of images we want to display. We apply `listdir` to access the subdirectories and to introduce randomness in the selection process. We use the `shuffle` function from the `random` library for shuffling and selecting images randomly. Matplotlib is used to display the specified number of random images in our function:

```
def view_random_images(target_dir, num_images):
    """

    View num_images random images from the subdirectories of
    target_dir as a subplot.
    """
    # Get list of subdirectories
      subdirs = [d for d in os.listdir(
          target_dir) if os.path.isdir(os.path.join(
              target_dir, d))]

    # Select num_images random subdirectories
      random.shuffle(subdirs)
      selected_subdirs = subdirs[:num_images]

    # Create a subplot
      fig, axes = plt.subplots(1, num_images, figsize=(15,9))
      for i, subdir in enumerate(selected_subdirs):
        # Get list of images in subdirectory
          image_paths = [f for f in os.listdir(
              os.path.join(target_dir, subdir))]
```

```
        # Select a random image
        image_path = random.choice(image_paths)
        # Load image
        image = plt.imread(os.path.join(target_dir,
            subdir, image_path))
        # Display image in subplot
        axes[i].imshow(image)
        axes[i].axis("off")
        axes[i].set_title(subdir)
    print(f"Shape of image: {image.shape}")
    #width,height, colour chDNNels
    plt.show()
```

7. Let's try out the function by setting `num_images` to 4 and examine some data in our `train` directory:

    ```
    view_random_images(target_dir="/content/drive/MyDrive/weather
    dataset/train/", num_images=4)
    ```

    This returns four randomly selected images, as illustrated here:

 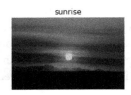 

7.18 – Randomly selected images from the weather dataset

From the data displayed, we can see the images come in various sizes (height and weight) and we will need to fix this preprocessing issue. We will be using the `ImageDataGenerator` class from TensorFlow. Let's discuss this next.

Image data preprocessing

We saw, in *Figure 7.18*, that our training images are of different sizes. Here, we will be resizing and normalizing our data before training. Also, we want to develop an efficient method of loading our training data in batches, ensuring optimized memory usage with seamless integration with our model's training process. To achieve all of this, we will be utilizing the `ImageDataGenerator` class from the `TensorFlow.keras.preprocessing.image` module. In *Chapter 8, Handling Overfitting,* we will take our application of `ImageDataGenerator` further by using it to enlarge our training dataset by developing variants of our image data by rotating, flipping, and zooming. This could help our model become more robust and reduce the risk of overfitting.

Another useful tool to aid our data preprocessing task is the `flow_from_directory` method. We can use this method to build data pipelines. It is especially useful when we are working on large-scale, real-world data because of its ability to automate reading, resizing, and batching images for model training or inference. The `flow_from_directory` method takes three main arguments. The first is the directory path that contains our image data. Next, we specify the desired size of the images before we feed them into our neural network. Then, we also have to specify the batch size to determine the number of images we want to process simultaneously. We can tailor the process more by specifying other parameters, such as color mode, class mode, and shuffle. Let's now take a look at a typical directory structure for a multiclass classification problem, as illustrated in *Figure 7.19*.

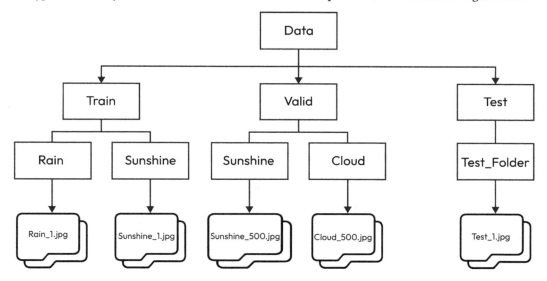

Figure 7.19 – The directory structure for a multiclass classification problem

When applying the `flow_from_directory` method, it is important that we organize our images in a well-structured directory, with subdirectories for each unique class label as displayed in *Figure 7.19*. Here, we have four subdirectories, one for each class label in our weather dataset. Once all the images are in the appropriate subdirectories, we can apply `flow_from_directory` to set up an iterator. This iterator is adjustable so that we can define parameters such as the image size and batch size and decide whether we want to shuffle our data or not. Let's apply these new ideas to our current case study:

```
# Preprocess data (get all of the pixel values between 1 and 0, also
called scaling/normalization)
train_datagen = ImageDataGenerator(rescale=1./255)
valid_datagen = ImageDataGenerator(rescale=1./255)
test_datagen = ImageDataGenerator(rescale=1./255)
```

Here, we define three instances of the `ImageDataGenerator` class: one for training, one for validation, and one for testing. We apply a rescaling factor of 1/255 to the pixel values of images in each instance to normalize our data:

```
# Import data from directories and turn it into batches
train_data = train_datagen.flow_from_directory(train_dir,
    batch_size=64, # number of images to process at a time
    target_size=(224,224), # convert all images to be 224 x 224
    class_mode="categorical")

valid_data = valid_datagen.flow_from_directory(val_dir,
    batch_size=64,
    target_size=(224,224),
    class_mode="categorical")
test_data = test_datagen.flow_from_directory(test_dir,
    batch_size=64,
    target_size=(224,224),
    class_mode="categorical")
```

We use `flow_from_directory` to import images from the respective training, validation, and testing directories, and the resulting data is stored in our `train_data`, `valid_data`, and `test_data` variables. In addition to specifying the directories in our `flow_from_directory` method, you will notice we also specified not just the target size (224 x244) and batch size (64) but also the type of problem we are tackling as `categorical` because we are dealing with a multi-classification use case. We have now successfully completed our data preprocessing steps. Let's move on to modeling our data:

```
model_1 = tf.keras.models.Sequential([
    tf.keras.layers.Conv2D(filters=16,
        kernel_size=3, # can also be (3, 3)
        activation="relu",
        input_shape=(224, 224, 3)),
        #(height, width, colour channels)
tf.keras.layers.MaxPool2D(2,2),
    tf.keras.layers.Conv2D(32, 3, activation="relu"),
    tf.keras.layers.MaxPool2D(2,2),
    tf.keras.layers.Conv2D(64, 3, activation="relu"),
    tf.keras.layers.MaxPool2D(2,2),
    tf.keras.layers.Flatten(),
    tf.keras.layers.Dense(1050, activation="relu"),
    tf.keras.layers.Dense(4, activation="softmax")
```

```
        # binary activation output
])

# Compile the model
model_1.compile(loss="CategoricalCrossentropy",
    optimizer=tf.keras.optimizers.Adam(),
    metrics=["accuracy"])

# Fit the model
history_1 = model_1.fit(train_data,
    epochs=10,
    validation_data=valid_data,
    )
```

Here, we use a CNN architecture made up of three sets of convolutional and pooling layers. In the first convolutional layer, we apply 16 filters with a filter size of 3. Notice the input shape also matches the shape defined in our preprocessing step. After the first convolutional layer, we apply max pooling of 2x2. Next, we reach the second convolutional layer, which utilizes 32 filters, each 3x3 in size, followed by another 2x2 max pooling layer. The final convolutional layer has 64 filters, each 3x3 in size, followed by another max pooling layer, which further downsamples the data.

Next, we reach the fully connected layers. Here, we first flatten the 3D output of the earlier layers into a 1D array. Then, we feed the data into dense layers for final classification. We proceed by compiling and fitting our model to our data. It's important to note that in our compile step, we use CategoricalCrossentropy for our loss function as we are dealing with a task with multiple classes, and we set metrics to accuracy. The resulting output is a probability distribution over the four classes in our dataset, with the class with the highest probability being the predicted label:

```
Epoch 6/10
13/13 [==============================] - 8s 622ms/step - loss: 0.1961
- accuracy: 0.9368 - val_loss: 0.2428 - val_accuracy: 0.8994
Epoch 7/10
13/13 [==============================] - 8s 653ms/step - loss: 0.1897
- accuracy: 0.9241 - val_loss: 0.2967 - val_accuracy: 0.9218
Epoch 8/10
13/13 [==============================] - 8s 613ms/step - loss: 0.1093
- accuracy: 0.9671 - val_loss: 0.3447 - val_accuracy: 0.8939
Epoch 9/10
13/13 [==============================] - 8s 604ms/step - loss: 0.1756
- accuracy: 0.9381 - val_loss: 0.6276 - val_accuracy: 0.8324
Epoch 10/10
13/13 [==============================] - 8s 629ms/step - loss: 0.1472
- accuracy: 0.9418 - val_loss: 0.2633 - val_accuracy: 0.9106
```

We train our model for 10 epochs, attaining a training accuracy of around 94% on training data and 91% on validation data. We use the `summary` method to obtain information about the different layers in the model. This information includes the layer-wise overview, output shape, and number of parameters used (trainable and non-trainable):

```
Model: "sequential"

_____
 Layer (type)                Output Shape              Param #
=================================================================
 conv2d (Conv2D)             (None, 222, 222, 16)  448

 max_pooling2d (MaxPooling2D) (None, 111, 111, 16)  0

 conv2d_1 (Conv2D)           (None, 109, 109, 32)  4640

 max_pooling2d_1 (MaxPooling  (None, 54, 54, 32)    0
 2D)

 conv2d_2 (Conv2D)           (None, 52, 52, 64)    18496

 max_pooling2d_2 (MaxPooling  (None, 26, 26, 64)    0
 2D)

 flatten (Flatten)           (None, 43264)          0

 dense (Dense)               (None, 1050)           45428250

 dense_1 (Dense)             (None, 4)              4204

=================================================================
Total params: 45,456,038
Trainable params: 45,456,038
Non-trainable params: 0
```

From our model's summary, we see our architecture has three convolutional (`Conv2D`) layers, each accompanied by a pooling (`MaxPooling2D`) layer. Information flows from these layers into the fully connected layer, where the final classification is carried out. Let's drill down into each of the layers and unpack the information they provide us with. The first convolutional layer is with an output shape of (`None, 222, 222, 16`). Here, `None` means we didn't hardcode the batch size, which gives us the flexibility to use different batch sizes with ease. Next, we have `222, 222`, which represents the dimension of the output feature map; we lose 2 pixels in height and weight because of the boundary effect if we do not apply padding. Finally, `16` represents the number of filters or kernels used, which means we will have an output of 16 different feature maps from each of the filters. You will also notice this layer has 448 parameters. To calculate the number of parameters in the convolutional layers, we use the following formula:

(Filter width × Filter height × Input channels + 1(for bias)) × Number of filters = Total number of parameters in the convolutional layer

When we key in the values into the formula, we arrive at $(3 \times 3 \times 3 + 1) \times 16 = 448$ parameters.

The next layer is the first pooling layer, which is a `MaxPooling2D` layer that downsamples the output feature maps from the convolutional layer. Here, we have an output shape of (`None, 111, 111, 16`). From the output, you can see that the spatial dimension has been reduced to half, and it is also important to note that pooling layers have no parameters, as you will observe with all the pooling layers in our model's summary.

Next, we reach the second convolutional layer and notice the depth of our output has increased to 32. This happens because we employed 32 filters in this layer; hence, we will have 32 different feature maps returned. Also, we have the spatial dimension of the feature maps again reduced by two pixels because of the boundary effect. We can easily calculate the number of parameters in this layer as follows: $(3 \times 3 \times 16 + 1) \times 32 = 4,640$ parameters.

Next, we reach the second pooling layer, which downsamples the feature maps further to (`None, 54, 54, 32`). The final convolutional layer uses 64 filters, so it has an output shape of (`None, 52, 52, 64`) and 18,496 parameters. The final pooling layer again reduces the dimension of our data to (`None, 26, 26, 64`). The output of the final pooling layer is fed into the `Flatten` layer, which reshapes the data from a 3D tensor into a 1D tensor with a size of 26 x 26 x 64 = 43,264. This is fed into the first `Dense` layer, which has an output shape of (`None, 1050`). To calculate the number of parameters in the `Dense` layer, we use this formula:

(Number of input nodes + 1) × Number of output nodes

When we input the values, we get $(43,264 + 1) \times 1,050 = 45,428,250$ parameters. The final `Dense` layer is the output layer and it has a shape of (`None, 4`), where 4 represents the number of unique classes in our data that we want to predict. This layer has $(1,050 + 1) \times 4 = 4,204$ parameters due to its connections, biases, and the number of output neurons.

Next, we evaluate our model using the `evaluate` method:

```
model_1.evaluate(test_data)
```

We reach an accuracy of 91% on our test data.

Let's compare our CNN architecture with two DNNs:

```
model_2 = tf.keras.Sequential([
    tf.keras.layers.Flatten(input_shape=(224, 224, 3)),
    tf.keras.layers.Dense(1200, activation='relu'),
```

```
        tf.keras.layers.Dense(600, activation='relu'),
        tf.keras.layers.Dense(300, activation='relu'),
        tf.keras.layers.Dense(4, activation='softmax')
])

# Compile the model
model_2.compile(loss='categorical_crossentropy',
        optimizer=tf.keras.optimizers.Adam(),
        metrics=["accuracy"])

# Fit the model
history_2 = model_2.fit(train_data,
        epochs=10,
        validation_data=valid_data)
```

We build a DNN called model_2 made up of 4 Dense layers, with 1200, 600, 300, and 4 neurons, respectively. Apart from the output layer, which uses the softmax function for classification, all the other layers use ReLU as their activation function. We compile and fit model_2 in the same way as model_1:

```
Epoch 6/10
13/13 [==============================] - 8s 625ms/step - loss: 2.2083
- accuracy: 0.6953 - val_loss: 0.9884 - val_accuracy: 0.7933
Epoch 7/10
13/13 [==============================] - 8s 606ms/step - loss: 2.7116
- accuracy: 0.6435 - val_loss: 2.0749 - val_accuracy: 0.6704
Epoch 8/10
13/13 [==============================] - 8s 636ms/step - loss: 2.8324
- accuracy: 0.6877 - val_loss: 1.7241 - val_accuracy: 0.7430
Epoch 9/10
13/13 [==============================] - 8s 599ms/step - loss: 1.8597
- accuracy: 0.6890 - val_loss: 1.1507 - val_accuracy: 0.7877
Epoch 10/10
13/13 [==============================] - 8s 612ms/step - loss: 1.0902
- accuracy: 0.7813 - val_loss: 0.9915 - val_accuracy: 0.7486
```

After 10 epochs, we reach a validation accuracy of 74.86% and when we examine the model's summary, we see that we have used a total of 181,536,904 parameters, which is 4 times the size of our CNN architecture parameters.

Next, let's look at another DNN architecture:

```
model_3 = tf.keras.Sequential([
        tf.keras.layers.Flatten(input_shape=(224, 224, 3)),
        tf.keras.layers.Dense(1000, activation='relu'),
        tf.keras.layers.Dense(500, activation='relu'),
```

```
        tf.keras.layers.Dense(500, activation='relu'),
        tf.keras.layers.Dense(4, activation='softmax')
])


# Compile the model
model_3.compile(loss='categorical_crossentropy',
    optimizer=tf.keras.optimizers.Adam(),
    metrics=["accuracy"])


# Fit the model
history_3 = model_3.fit(train_data,
    epochs=10,
    validation_data=valid_data)
```

We use another set of 4 Dense layers, with 1000, 500, 500, and 4 neurons, respectively. We fit and compile model_3 as well for 10 epochs:

```
Epoch 6/10
13/13 [==============================] - 9s 665ms/step - loss: 1.6911
- accuracy: 0.6814 - val_loss: 0.5861 - val_accuracy: 0.7877
Epoch 7/10
13/13 [==============================] - 8s 606ms/step - loss: 0.7309
- accuracy: 0.7952 - val_loss: 0.5100 - val_accuracy: 0.8268
Epoch 8/10
13/13 [==============================] - 8s 572ms/step - loss: 0.6797
- accuracy: 0.7863 - val_loss: 0.9520 - val_accuracy: 0.7263
Epoch 9/10
13/13 [==============================] - 8s 632ms/step - loss: 0.7430
- accuracy: 0.7724 - val_loss: 0.5220 - val_accuracy: 0.7933
Epoch 10/10
13/13 [==============================] - 8s 620ms/step - loss: 0.5845
- accuracy: 0.7737 - val_loss: 0.5881 - val_accuracy: 0.7765
```

We reach a validation accuracy of 77.65% after 10 epochs and this model has around 151,282,004 parameters; the results are not close to those of our CNN architecture. Let's proceed to compare all three models on test data, which is what we want to be judging our models on. To do this, we will write a function to generate a DataFrame showing the names, the loss, and the accuracy of the models:

```
def evaluate_models(models, model_names, test_data):
    # Initialize lists for the results
    losses = []
    accuracies = []

    # Iterate over the models
    for model in models:
        # Evaluate the model
```

```
        loss, accuracy = model.evaluate(test_data)
        losses.append(loss)
        accuracies.append(accuracy)
    # Convert the results to percentages
    losses = [round(loss * 100, 2) for loss in losses]
    accuracies = [round(accuracy * 100, 2) for accuracy in accuracies]

    # Create a dataframe with the results
    results = pd.DataFrame({"Model": model_names,
        "Loss": losses,
        "Accuracy": accuracies})

    return results
```

The `evaluate_models()` function takes a list of models, model names, and test data as input and returns a DataFrame with the evaluation results for each model as percentages:

```
# Define the models and model names
models = [model_1, model_2, model_3]
model_names = ["Model 1", "Model 2", "Model 3"]

# Evaluate the models
results = evaluate_models(models, model_names,test_data)

# Display the results
results
```

When we run the code, it generates the table shown in *Figure 7.20*.

	Model	Loss	Accuracy
0	Model 1	26.33	91.06
1	Model 2	99.15	74.86
2	Model 3	58.81	77.65

Figure 7.20 – A DataFrame showing the experimental results of all three models

From the results, we can clearly see that Model 1 is ahead. You may wish to experiment with larger DNNs but you will soon run out of memory. For larger datasets, the results may be much worse for DNNs. Next, let's look at how we fared on our training and validation data with Model 1:

```
def plot_loss_accuracy(history_1):
    # Extract the loss and accuracy history for both training and
validation data
    loss = history_1.history['loss']
    val_loss = history_1.history['val_loss']
    acc = history_1.history['accuracy']
    val_acc = history_1.history['val_accuracy']

    # Create subplots
    fig, (ax1, ax2) = plt.subplots(1, 2, figsize=(9, 6))

    # Plot the loss history
    ax1.plot(loss, label='Training loss')
    ax1.plot(val_loss, label='Validation loss')
    ax1.set_title('Loss history')
    ax1.set_xlabel('Epoch')
    ax1.set_ylabel('Loss')
    ax1.legend()

    # Plot the accuracy history
    ax2.plot(acc, label='Training accuracy')
    ax2.plot(val_acc, label='Validation accuracy')
    ax2.set_title('Accuracy history')
    ax2.set_xlabel('Epoch')
    ax2.set_ylabel('Accuracy')
    ax2.legend()

    plt.show()
```

We created a function to plot the training and validation loss and accuracy using matplotlib. We pass history_1 into our function:

```
# Lets plot the training and validation loss and accuracy
plot_loss_accuracy(history_1)
```

This will generate the following output:

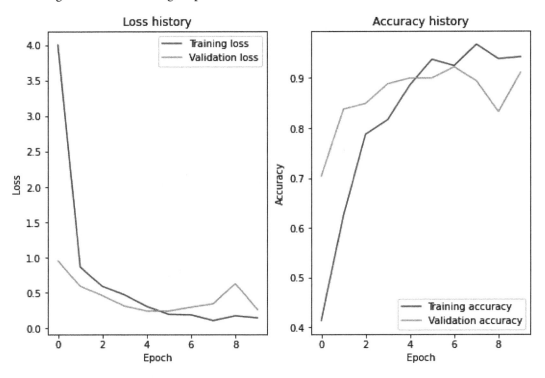

Figure 7.21 – Loss and accuracy plot for Model 1

From the plot, we can see that our training accuracy rises steadily but falls below its highest point just before the 10th epoch. Also, our validation data experiences a sharp fall in accuracy. Our loss veers off from the fourth epoch.

Summary

In this chapter, we saw the power of CNNs. We began by examining the challenges faced by DNNs for visual recognition tasks. Next, we journeyed through the anatomy of CNNs, zooming in on the various moving parts, such as the convolutional, pooling, and fully connected layers. Here, we saw the impact and effect of different hyperparameters, and we also discussed the boundary effect. Next, we moved on to using all we learned to build a real-world weather classifier using two DNNs and a CNN. Our CNN model outperformed the DNNs, showcasing the strength of CNNs in handling image-based problems. Also, we discussed and applied some TensorFlow functions that streamline data preprocessing and modeling when we are working with image data.

By now you should have a good understanding of the structure and operations of CNNs and how to use them to solve real-world image classification problems, as well as utilizing various tools in TensorFlow to effectively and efficiently preprocess image data for improved model performance. In the next chapter, we will address the issue of overfitting in neural networks and explore various techniques to overcome this challenge, ensuring that our models generalize well to unseen data.

In the next chapter, we will use some old tricks such as callbacks and hyperparameter tuning to see whether we can improve our model's performance. We will also experiment with data augmentation and other new techniques to improve our model's performance. We draw the curtains on our task for now until *Chapter 8, Handling Overfitting*.

Questions

Let's test what we have learned in this chapter:

1. What are the components of a typical CNN architecture?

2. How does a convolutional layer work in a CNN architecture?

3. What is pooling and why is it used in a CNN architecture?

4. What is the purpose of a fully connected layer in a CNN architecture?

5. What is the impact of the padding on a convolution operation?

6. What are the advantages of using TensorFlow image data generators?

Further reading

To learn more, you can check out the following resources:

- Dumoulin, V., & Visin, F. (2016). *A guide to convolution arithmetic for deep learning.* http://arxiv.org/abs/1603.07285

- Gulli, A., Kapoor, A. and Pal, S., 2019. *Deep Learning with TensorFlow 2 and Keras.* Birmingham: Packt Publishing Ltd

- Kapoor, A., Gulli, A. and Pal, S. (2020) *Deep Learning with TensorFlow and Keras Third Edition: Build and deploy supervised, unsupervised, deep, and reinforcement learning models.* Packt Publishing Ltd

- Krizhevsky, A., Sutskever, I., & Hinton, G. E. (2012). *ImageNet classification with deep convolutional neural networks.* In Advances in neural information processing systems (pp. 1097-1105)

- Zhang, Y., & Yang, H. (2018). *Food classification with convolutional neural networks and multi-class linear discernment analysis.* In 2018 IEEE International Conference on Information Reuse and Integration (IRI) (pp. 1-5). IEEE

- Zhang, Z., Ma, H., Fu, H., & Zha, C. (2020). *Scene-Free Multi-Class Weather Classification on Single Images*. IEEE Access, 8, 146038-146049. doi:10.1109

- The `tf.image` module: `https://www.tensorflow.org/api_docs/python/tf/image`

8
Handling Overfitting

One major challenge in **machine learning** (**ML**) is overfitting. **Overfitting** occurs when a model is trained too well on the training data but fails to generalize on unseen data, resulting in poor performance. In *Chapter 6, Improving the Model* we witnessed firsthand how overtraining pushed our model into this overfitting trap. In this chapter, we will probe further into the nuances of overfitting, striving to unpack both its warning signs and the underlying reasons behind it. Also, we will explore the different strategies we can apply to mitigate the dangers overfitting presents to real-world ML applications. Using TensorFlow, we will apply these ideas in a hands-on fashion to overcome overfitting in a real-world case study. By the end of this chapter, you should have a solid understanding of what overfitting is and how to mitigate it in real-world image classification tasks.

In this chapter, we will cover the following topics:

- Overfitting in ML
- Early stopping
- Changing the model's architecture
- L1 and L2 regularization
- Dropout regularization
- Data augmentation

Technical requirements

We will be using Google Colab to run the coding exercise that requires `python >= 3.8.0`, along with the following packages, which can be installed using the `pip install` command:

- `tensorflow>=2.7.0`
- `os`
- `pillow==8.4.0`

- `pandas==1.3.4`

- `numpy==1.21.4`

- `matplotlib >=3.4.0`

The code bundle for this book is available at the following GitHub link: `https://github.com/PacktPublishing/TensorFlow-Developer-Certificate`. Also, solutions to all exercises can be found in the GitHub repo itself.

Overfitting in ML

From the previous chapters, we now know what overfitting is and its adverse effect when used on unseen data. Let's take a step further by digging into what the root causes of overfitting are, how we can spot overfitting when we build our models, and some important strategies we can apply to curb overfitting. When we gain this understanding, we can go on to build effective and robust ML models.

What triggers overfitting

In *Chapter 6, Improving the Model,* we saw that by adding more neurons to our hidden layer, our model became too complex. This made our model not only capture the patterns in our data but also the noise in it, leading to overfitting. Another root cause of overfitting is working with insufficient data volume. If our data does not truly capture the full spectrum of variations our model will be faced with upon deployment, when we train our model on such a dataset, it becomes too specialized and fails to generalize when used in the real world.

Beyond the volume of data, another issue we can face is noisy data. Unlike when we work with curated or static data, when building real-world applications, we may find that our data could be noisy or incorrect. If we develop models with such data, there is a chance it would lead to overfitting when put to use. We looked at some ideas around why overfitting could occur; the next question we may want to ask is, how can we detect overfitting? Let's discuss this in the following subsection.

Detecting overfitting

One way we can detect overfitting is by comparing a model's accuracy on training data versus the validation/test data. When a model records a high accuracy on training and poor accuracy on testing, this disparity indicates that the model has memorized the training samples, hence its poor generalization of unseen data. Another effective way of discovering overfitting is to examine the training error against the validation error. When the training error decreases over time but the validation error increases, this can indicate our model overfits, as the model performs worse on the validation data. A scenario where the model's validation accuracy deteriorates as its training counterpart flourishes should sound the alarm bells for potential overfitting.

Let's revisit our case study from *Chapter 7, Image Classification with Convolutional Neural Networks,* the weather dataset from WeatherBIG, and examine how we can monitor overfitting by using a validation dataset during the model's training process. By employing a validation dataset, we can accurately track the model's performance and prevent overfitting. Let's begin by creating a baseline model.

Baseline model

Following the standard three-step approach of building, compiling, and fitting, we will construct a **convolutional neural network (CNN)** model comprising two Conv2D and pooling layers, coupled with a fully connected layer that has a dense layer of 1,050 neurons. The output layer consists of four neurons, which represent the four classes in our dataset. We then compile and fit the model using the training data for 20 epochs:

```
#Build
model_1 = tf.keras.models.Sequential([
    tf.keras.layers.Conv2D(filters=16,
        kernel_size=3, # can also be (3, 3)
        activation="relu",
        input_shape=(224, 224, 3)),
        #(height, width, colour channels)
    tf.keras.layers.MaxPool2D(2,2),
    tf.keras.layers.Conv2D(32, 3, activation="relu"),
    tf.keras.layers.MaxPool2D(2,2),
    tf.keras.layers.Conv2D(64, 3, activation="relu"),
    tf.keras.layers.MaxPool2D(2,2),
    tf.keras.layers.Flatten(),
    tf.keras.layers.Dense(1050, activation="relu"),
    tf.keras.layers.Dense(4, activation="softmax")
])


# Compile the model
model_1.compile(loss="CategoricalCrossentropy",
    optimizer=tf.keras.optimizers.Adam(),
    metrics=["accuracy"])


#fit
history_1 = model_1.fit(train_data,
    epochs=20,
    validation_data=valid_data
)
```

We set the `validation_data` parameter to `valid_data`. This ensures that when we run the code, after each epoch, the model will evaluate its performance on the validation data, as shown in *Figure 8.1*.

```
Epoch 26/30
25/25 [==============================] - 8s 315ms/step - loss: 3.9823e-04 - accuracy: 1.0000 - val_loss: 0.4997 - val_accuracy: 0.8994
Epoch 27/30
25/25 [==============================] - 8s 313ms/step - loss: 3.3314e-04 - accuracy: 1.0000 - val_loss: 0.5290 - val_accuracy: 0.8939
Epoch 28/30
25/25 [==============================] - 8s 312ms/step - loss: 3.0940e-04 - accuracy: 1.0000 - val_loss: 0.4979 - val_accuracy: 0.9050
Epoch 29/30
25/25 [==============================] - 8s 304ms/step - loss: 2.5297e-04 - accuracy: 1.0000 - val_loss: 0.5163 - val_accuracy: 0.9050
Epoch 30/30
25/25 [==============================] - 8s 316ms/step - loss: 2.5945e-04 - accuracy: 1.0000 - val_loss: 0.4891 - val_accuracy: 0.9106
```

Figure 8.1 – The last five training epochs

This is a straightforward way to compare the loss values between the training set and the validation set. We can see that the model accurately predicts every sample in the training set, reaching an accuracy of 100 percent. However, on the validation set, it attains an accuracy of 91 percent, which suggests that the model likely overfits. Another effective way to observe overfitting is to use the learning curve to plot the loss and accuracy values of both the training set and the validation set – again, a large gap between both plots is a sign of overfitting, as shown in *Figure 8.2*.

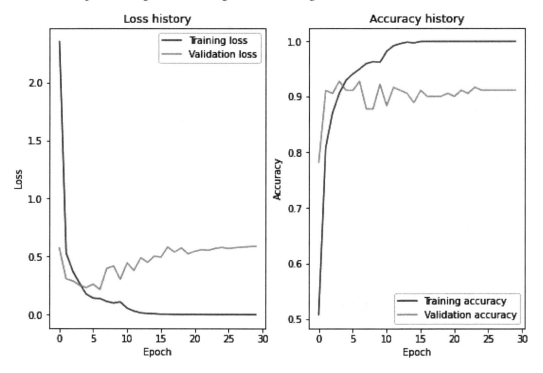

Figure 8.2 – The learning curve showing the loss and accuracy for both training and test data

At the start of the experiment, the difference between the training loss and the validation loss is minimal; however, as we move into the fourth epoch, the validation loss starts to increase, while the training loss continues to decrease. Similarly, the training and validation accuracies are closely aligned at the start, but again, at around the fourth epoch, the validation accuracy reaches a peak of around 90 percent and stays there, while the training accuracy reaches 100 percent accuracy.

The ultimate objective of building an image classifier is to apply it to real-world data. After completing the training process, we evaluate the model on our holdout dataset. If the results obtained during testing are significantly different from those achieved during training, this could indicate overfitting.

Fortunately, there are several strategies that can be applied to overcome overfitting. Some of the main techniques to handle overfitting focus on improving the model itself to enhance its generalization capabilities. On the other hand, it is equally important to examine the data itself, observing what the model missed during training and evaluation. By visualizing the misclassified images, we gain insight into where the model falls short. We start by first recreating our baseline model from *Chapter 7, Image Classification with Convolutional Neural Networks*. This time, we train it for 20 epochs to enable us to observe overfitting, as illustrated in *Figure 8.2*. Next, let's see how we can curb overfitting using several strategies, starting with applying early stopping.

Early stopping

In *Chapter 6, Improving the Model*, we introduced the concept of early stopping as an effective way of preventing overfitting. It does this by halting training when the model's performance fails to improve over a defined number of epochs, as indicated in *Figure 8.3*. This way, we prevent our model from overfitting.

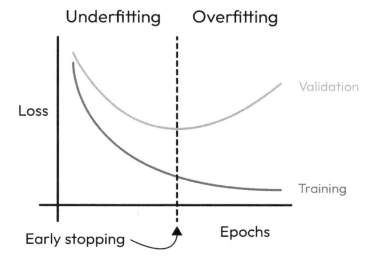

Figure 8.3 – A learning curve showing early stopping

Let's recreate the same baseline model, but this time, we will apply a built-in callback to stop training when the validation accuracy fails to improve. We will use the same build and compile steps as in the first model and then add a callback when we fit the model:

```
#Fit the model
# Add an early stopping callback
callbacks = [tf.keras.callbacks.EarlyStopping(
    monitor="val_accuracy", patience=3,
    restore_best_weights=True)]
history_2 = model_2.fit(train_data,
    epochs=20,
    validation_data=valid_data,
    callbacks=[callbacks])
```

Here, we specified the number of epochs as 20 and added a validation set to monitor the model's performance during training. After this, we used the callbacks argument to specify a callback function to implement early stopping. We used an early stopping callback to stop training after three epochs should the validation set accuracy fail to improve. This is done by setting the patience parameter to 3. This means that if there's no progress in validation accuracy for three straight epochs, the early stopping callback halts training. We also set the restore_best_weights parameter to True; this restores the best model weight from the training process when the training ends. The information from the fit function is stored in the history_2 variable:

```
Epoch 8/20
25/25 [==============================] - 8s 318ms/step - loss: 0.0685
- accuracy: 0.9810 - val_loss: 0.3937 - val_accuracy: 0.8827
Epoch 9/20
25/25 [==============================] - 8s 325ms/step - loss: 0.0368
- accuracy: 0.9912 - val_loss: 0.3338 - val_accuracy: 0.9218
Epoch 10/20
25/25 [==============================] - 8s 316ms/step - loss: 0.0169
- accuracy: 0.9987 - val_loss: 0.4322 - val_accuracy: 0.8994
Epoch 11/20
25/25 [==============================] - 8s 297ms/step - loss: 0.0342
- accuracy: 0.9912 - val_loss: 0.2994 - val_accuracy: 0.8994
Epoch 12/20
25/25 [==============================] - 8s 318ms/step - loss: 0.1352
- accuracy: 0.9570 - val_loss: 0.4503 - val_accuracy: 0.8939
```

From the training process, we can see that our model reaches a peak validation accuracy of 0.9218 on the ninth epoch, after which training continues for three epochs before stopping. Since there were no further improvements in the validation accuracy, training is stopped and the best weight is saved. Now, let's evaluate model_2 on our test data:

```
model_2.evaluate(test_data)
```

When we run the code, we see that the model achieves an accuracy of 0.9355. Here, the performance on the test set is in line with the performance on the validation set and higher than our baseline model, where we achieved an accuracy of 0.9097. This is our first step to create a better model.

```
Layer (type)                      Output Shape            Param #
=================================================================
conv2d (Conv2D)                   (None, 222, 222, 16)    448

max_pooling2d (MaxPooling2D       (None, 111, 111, 16)    0
)

conv2d_1 (Conv2D)                 (None, 109, 109, 32)    4640

max_pooling2d_1 (MaxPooling       (None, 54, 54, 32)      0
2D)

conv2d_2 (Conv2D)                 (None, 52, 52, 64)      18496

max_pooling2d_2 (MaxPooling       (None, 26, 26, 64)      0
2D)

flatten (Flatten)                 (None, 43264)           0

dense (Dense)                     (None, 1050)            45428250

dense_1 (Dense)                   (None, 4)               4204

=================================================================
Total params: 45,456,038
Trainable params: 45,456,038
Non-trainable params: 0
```

Figure 8.4 – A snapshot of the model summary

When we inspect our model summary, we can see that our model has over 45 million parameters, and this could lead to the model being susceptible to picking up noise in the training data, due to the model being highly parameterized. To address this issue, we can simplify our model by reducing the number of parameters in such a way that our model is not too complex for our dataset. Let's discuss model simplification next.

Model simplification

To address overfitting, you may consider reassessing the model's architecture. Simplifying your model's architecture could prove to be an effective strategy in tackling overfitting, especially when your model is highly parameterized. However, it is important to know that this approach does not always guarantee better performance in every instance; in fact, you must be mindful of oversimplifying your model, which could lead to the trap of underfitting. Hence, it is important to strike the right balance between model complexity and simplicity to achieve an optimally performing model, as illustrated in *Figure 8.5*, as the relationship between model complexity and overfitting is not a linear one.

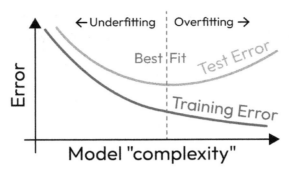

Figure 8.5 – Overfitting and underfitting in ML

Model simplification can be achieved in a number of ways – for instance, we can replace a large number of filters with smaller ones, or we could also reduce the number of neurons in the first Dense layer. In our architecture, you can see that the first dense layer has 1050 neurons. Let's reduce the neurons to 500 as the initial step in our model simplification experiment:

```
    tf.keras.layers.Flatten(),
    tf.keras.layers.Dense(500, activation="relu"),
    tf.keras.layers.Dense(4, activation="softmax")
])
```

When we compile and fit the model, our model reaches a peak accuracy of 0.9162 on the validation set:

```
Epoch 5/50
25/25 [==============================] - 8s 300ms/step - loss: 0.1284
- accuracy: 0.9482 - val_loss: 0.4489 - val_accuracy: 0.8771
Epoch 6/50
25/25 [==============================] - 8s 315ms/step - loss: 0.1122
- accuracy: 0.9659 - val_loss: 0.2414 - val_accuracy: 0.9162
Epoch 7/50
25/25 [==============================] - 8s 327ms/step - loss: 0.0814
- accuracy: 0.9735 - val_loss: 0.2976 - val_accuracy: 0.9050
Epoch 8/50
25/25 [==============================] - 11s 441ms/step - loss: 0.0541
- accuracy: 0.9785 - val_loss: 0.2215 - val_accuracy: 0.9050
Epoch 9/50
25/25 [==============================] - 8s 313ms/step - loss: 0.1279
- accuracy: 0.9621 - val_loss: 0.2848 - val_accuracy: 0.8994
```

Since our validation accuracy did not fare better, perhaps now will be a good time to try a few well-known ideas to fix overfitting. Let's look at L1 and L2 regularizations in the following subsection. We will discuss how they work and apply them to our case study.

> **Note**
> The goal of model simplification is not to achieve a smaller model but a well-designed model that generalizes well. We may just need to reduce layers if they are unnecessary for our use case, or we could simplify the model by changing the activation function or reorganizing the order and arrangement of the model layers to improve the flow of information.

L1 and L2 regularization

Regularization is a set of techniques used to prevent overfitting by reducing a model's complexity, by applying a penalty term to the loss function. Regularization techniques make the model more resistant to the noise in the training data, thus improving its ability to generalize to unseen data. There are different types of regularization techniques, namely L1 and L2 regularization. **L1 and L2 regularization** are two well know regularization techniques; L1 can also be referred to as **lasso regression**. When selecting between L1 and L2, it is important to consider the type of data we work with.

L1 regularization comes in handy when working with data with many irrelevant features. The penalty term in L1 will cause some of the coefficients to become zero, resulting in a reduction in the number of features used during modeling; this, in turn, reduces the risk of overfitting, as the model will be trained on less noisy data. Conversely, L2 is an excellent choice when the goal is to create a model with small weights and good generalization. The penalty term in L2 reduces the magnitude of the coefficients, preventing them from becoming too large and leading to overfitting:

```
tf.keras.layers.Flatten(),
tf.keras.layers.Dense(1050, activation="relu",
    kernel_regularizer=regularizers.l2(0.01)),
tf.keras.layers.Dense(4, activation="softmax")
])
```

When we run this experiment, we reach an accuracy of around 92 percent, not faring better than other experiments. To try out L1 regularization, we simply changed the regularization method from L2 to L1. However, in this case, our results were not as good. As a result, let's try another regularization method called dropout regularization.

Dropout regularization

One key issue with neural networks is co-dependence. **Co-dependence** is a phenomenon in neural networks that occurs when a group of neurons, especially in the same layer, become highly correlated such that they rely too much on each other. This could lead to them amplifying certain features and failing to capture other important features in the data. Because these neurons act in sync, our model is more prone to overfitting. To mitigate this risk, we can apply a technique referred to as **dropout**. Unlike L1 and L2 regularization, dropout does not add a penalty term, but as the name implies, we randomly "drop out" a certain percentage of neurons from the model during training, as illustrated in *Figure 8.6*, reducing co-dependence between neurons, which can help to mitigate against overfitting.

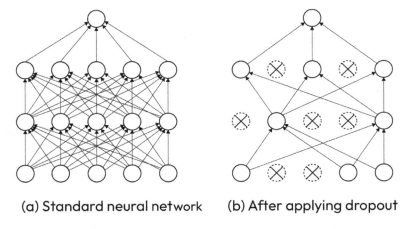

(a) Standard neural network   (b) After applying dropout

Figure 8.6 – A neural network with dropout applied

When we apply the dropout technique, the model is forced to learn more robust features, since we break co-dependence between neurons. However, it's worth noting that when we apply dropout, the training process may require more iterations to achieve convergence. Let's apply dropout to our baseline model and observe what its effect will be:

```
tf.keras.layers.Flatten(),
tf.keras.layers.Dense(1050, activation="relu"),
tf.keras.layers.Dropout(0.6), # added dropout layer
tf.keras.layers.Dense(4, activation="softmax")])
```

To implement dropout in code, we specify the dropout layer using the `tf.keras.layers.Dropout(0.6)` function. This creates a dropout layer with a dropout rate of `0.6` – that is, we turn off 60 percent of the neurons during training. It is worth noting that we can set the dropout value between 0 and 1:

```
25/25 [==============================] - 8s 333ms/step - loss: 0.3069
- accuracy: 0.8913 - val_loss: 0.2227 - val_accuracy: 0.9330
Epoch 6/10
25/25 [==============================] - 8s 317ms/step - loss: 0.3206
- accuracy: 0.8824 - val_loss: 0.1797 - val_accuracy: 0.9441
Epoch 7/10
25/25 [==============================] - 8s 322ms/step - loss: 0.2557
- accuracy: 0.9166 - val_loss: 0.2503 - val_accuracy: 0.8994
Epoch 8/10
25/25 [==============================] - 9s 339ms/step - loss: 0.1474
```

```
- accuracy: 0.9469 - val_loss: 0.2282 - val_accuracy: 0.9274
Epoch 9/10
25/25 [==============================] - 8s 326ms/step - loss: 0.2321
- accuracy: 0.9241 - val_loss: 0.3958 - val_accuracy: 0.8659
```

In this experiment, our model reaches a peak performance of 0.9441 on the validation set, improving our baseline model's performance. Next, let's look at changing the learning rate.

Adjusting the learning rate

In *Chapter 6, Improving the Model,* we discussed learning rates and the need to find an optimal learning rate. For this experiment, let us use a learning rate of 0.0001, which I found to produce a good result here, by experimenting with different learning rates, similar to what we did in *Chapter 6, Improving the Model.* In *Chapter 13, Time Series, Sequence and Prediction with TensorFlow,* we will look at how to apply both custom and inbuilt learning rate schedulers. Here, we also apply our early stopping callback to ensure that training is terminated once the model fails to improve. Let's compile our model:

```
# Compile the model
model_7.compile(loss="CategoricalCrossentropy",
    optimizer=tf.keras.optimizers.Adam(learning_rate=0.0001),
    metrics=["accuracy"])
```

We will fit the model and run it. In seven epochs, our model's training stops, reaching peak performance of 0.9274 on the validation set:

```
Epoch 3/10
25/25 [==============================] - 8s 321ms/step - loss: 0.4608
- accuracy: 0.8508 - val_loss: 0.2776 - val_accuracy: 0.8994
Epoch 4/10
25/25 [==============================] - 8s 305ms/step - loss: 0.3677
- accuracy: 0.8824 - val_loss: 0.2512 - val_accuracy: 0.9274
Epoch 5/10
25/25 [==============================] - 8s 316ms/step - loss: 0.3143
- accuracy: 0.8925 - val_loss: 0.4450 - val_accuracy: 0.8324
Epoch 6/10
25/25 [==============================] - 8s 317ms/step - loss: 0.2749
- accuracy: 0.9052 - val_loss: 0.3427 - val_accuracy: 0.8603
Epoch 7/10
25/25 [==============================] - 8s 322ms/step - loss: 0.2241
- accuracy: 0.9279 - val_loss: 0.2996 - val_accuracy: 0.8659
```

We've explored various methods to improve our model and overcome overfitting. Now, let's shift our focus to the dataset itself and examine how error analysis can be useful.

Error analysis

From our results so far, we can see that our model fails to misclassify some labels correctly. To further improve the generalization ability of our model, it is a good idea to examine the mistakes made by the model, with the underlying idea to uncover patterns in the misclassified data so that the insights we gain from looking at the misclassified labels can be used to improve the model's generalization capability. This technique is referred to as **error analysis**. To perform error analysis, we begin by identifying misclassified labels on the validation/test set. Next, we put these errors into groups – for example, we can make a group to blur images or images taken under poor lighting conditions.

Based on the insights gained from the collected errors, we may need to adjust our model architecture or tune our hyperparameters, especially when certain features are not captured by the model. Also, our error analysis step can also point us to the need to improve our data size and quality. One effective way of resolving this is by applying data augmentation, a well-known technique to enrich our data size and quality. Let's discuss data augmentation next and apply it to our case study.

Data augmentation

Image **data augmentation** is a technique used to increase the size and diversity of our training set by the application of various transformations, such as rotating, flipping, cropping, and scaling to create new, synthetic data, as illustrated in *Figure 8.7*. For many real-world applications, data collection can be a very expensive and time-consuming process; hence, data augmentation comes in quite handy. Data augmentation helps the model to learn more robust features rather than allowing the model to memorize features, thereby improving the model's generalization capabilities.

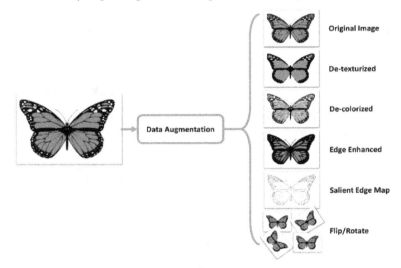

Figure 8.7 – Various data augmentation techniques applied to an image of a butterfly
(Source: https://medium.com/secure-and-private-ai-writing-challenge/data-augmentation-increases-accuracy-of-your-model-but-how-aa1913468722)

Another important use of data augmentation is to create balance across different classes in our training dataset. If the training set contains imbalanced data, we can use data augmentation techniques to create variants of the minority class, thereby building a more balanced dataset with a lower likelihood of overfitting. When implementing data augmentation, it's important to keep in mind various factors that may affect the outcome. For instance, the type of data augmentation to use depends on the type of data we work with.

In image classification tasks, techniques such as random rotations, translations, flips, and scaling may prove useful. However, when dealing with numeric datasets, applying rotations to numbers could lead to unintended results, such as rotating a 6 into a 9. Again, flipping letters of the alphabet, such as "b" and "d," can also have adverse effects. When applying image augmentation to our training set, it's crucial to consider the magnitude of augmentation and its effect on the quality of our training data. Excessive augmentation may lead to severely distorted images, resulting in a poorly performing model. To prevent this, it's equally important to monitor the model's training with a validation set.

Let's apply data augmentation to our case study and see what our results will look like.

To implement data augmentation, you can use the `ImageDataGenerator` class from the `tf.keras.preprocessing.image` module. This class allows you to specify a range of transformations, which should only be applied to images in our training set, and it generates synthetic images on the fly during the training process. For example, here is how you can use the `ImageDataGenerator` class to apply rotation, flipping, and scaling transformations to the training images:

```
train_datagen = ImageDataGenerator(rescale=1./255,
    rotation_range=25, zoom_range=0.3)
valid_datagen = ImageDataGenerator(rescale=1./255)

# Set up the train, validation, and test directories
train_dir = "/content/drive/MyDrive/weather dataset/train/"
val_dir = "/content/drive/MyDrive/weather dataset/validation/"
test_dir = "/content/drive/MyDrive/weather dataset/test/"

# Import data from directories and turn it into batches
train_data = train_datagen.flow_from_directory(
    train_dir,
    target_size=(224,224), # convert all images to be 224 x 224
    class_mode="categorical")

valid_data = valid_datagen.flow_from_directory(
    val_dir,
    target_size=(224,224),
    class_mode="categorical")
```

```
test_data = valid_datagen.flow_from_directory(
    test_dir,
    target_size=(224,224),
    class_mode="categorical",)
```

Using image data augmentation is quite straightforward; we created three instances of the ImageDataGenerator class from the keras.preprocessing.image module for our train, validation, and test sets. One key difference is that we added the rotation_range=25 and zoom_range=0.3 arguments to the train_datagen object. This will randomly rotate our images by 25 degrees and zoom them by a factor of 0.3 during the training process; everything else will remain the same.

Next, we will build, compile, and fit our baseline model, with early stopping applied, on our augmented data:

```
Epoch 4/20
25/25 [==============================] - 8s 308ms/step - loss: 0.2888
- accuracy: 0.9014 - val_loss: 0.3256 - val_accuracy: 0.8715
Epoch 5/20
25/25 [==============================] - 8s 312ms/step - loss: 0.2339
- accuracy: 0.9115 - val_loss: 0.2172 - val_accuracy: 0.9330
Epoch 6/20
25/25 [==============================] - 8s 320ms/step - loss: 0.1444
- accuracy: 0.9507 - val_loss: 0.2379 - val_accuracy: 0.9106
Epoch 7/20
25/25 [==============================] - 8s 315ms/step - loss: 0.1190
- accuracy: 0.9545 - val_loss: 0.2828 - val_accuracy: 0.9162
Epoch 8/20
25/25 [==============================] - 8s 317ms/step - loss: 0.0760
- accuracy: 0.9785 - val_loss: 0.3220 - val_accuracy: 0.8883
```

In eight epochs, our training comes to an end. This time, we reached 0.9330 on the validation set. So far, we have run seven different experiments. Let's test each of these models on the test set and examine what the results will look like. To do this, we will write a helper function that creates a DataFrame showing the top 5 models, each model's name, and the loss and accuracy of each model, as shown in *Figure 8.8*.

	Model	Loss	Accuracy
0	model 7	28.02	91.61
1	model 1	60.68	90.97
2	model 2	33.72	90.97
3	model 4	49.28	90.97
4	model 6	31.03	90.32

Figure 8.8 – A DataFrame showing the loss and accuracy of the top five models

Our best-performing model on our test data was **model 7**, where we altered the learning rate. We have covered a few ideas that are used to tackle overfitting in real-world image classifiers; however, a combination of these techniques can be applied to build a simpler yet more robust model that is less prone to overfitting. Combining various techniques of curbing overfitting is generally a good idea, as it may help to produce a more robust and generalizable model. However, it is important to keep in mind that there is no one-size-fits-all solution, and the best combination of methods will depend on the specific data and task at hand and may require multiple experiments.

Summary

In this chapter, we discussed overfitting in image classification and explored the different techniques to overcome it. We started by examining what overfitting is and why it happens, and we discussed how we can apply different techniques such as early stopping, model simplification, L1 and L2 regularization, dropout, and data augmentation to mitigate against overfitting in image classification tasks. Furthermore, we applied each of these techniques in our weather dataset case study and saw, hands-on, the effects of these techniques on our case study. We also explored combining these techniques in a quest to build an optimal model. By now, you should have a good understanding of overfitting and how to mitigate it in your own image classification projects.

In the next chapter, we will dive into transfer learning, a powerful technique that allows you to leverage pre-trained models for your specific image classification tasks, saving time and resources while achieving impressive results.

Questions

Let's test what we learned in this chapter:

1. What is overfitting in image classification tasks?
2. How does overfitting occur?
3. What techniques can be used to prevent overfitting?
4. What is data augmentation, and how is it used to prevent overfitting?
5. How can data pre-processing, data diversity, and data balancing be used to mitigate overfitting?

Further reading

To learn more, you can check out the following resources:

- Garbin, C., Zhu, X., & Marques, O. (2020). *Dropout vs. Batch Normalization: An Empirical Study of Their Impact to Deep Learning.* arXiv preprint arXiv:1911.12677: `https://par.nsf.gov/servlets/purl/10166570`.

- Kandel, I., & Castelli, M. (2020). *The effect of batch size on the generalizability of the convolutional neural networks on a histopathology dataset.* arXiv preprint arXiv:2003.00204.

- *Effect_batch_size_generalizability_convolutional_neural_networks_histopathology_dataset.pdf (unl.pt).* Kapoor, A., Gulli, A. and Pal, S. (2020): `https://research.unl.pt/ws/portalfiles/portal/18415506/Effect_batch_size_generalizability_convolutional_neural_networks_histopathology_dataset.pdf`.

- *Deep Learning with TensorFlow and Keras, Third Edition, Amita Kapoor, Antonio Gulli, Sujit Pal,* Packt Publishing Ltd.

- Nitish Srivastava, Geoffrey Hinton, Alex Krizhevsky, Ilya Sutskever, and Ruslan Salakhutdinov. 2014. *Dropout: A simple way to prevent neural networks from overfitting.* J. Mach. Learn. Res. 15, 1 (2014), 1,929–1,958 `https://jmlr.org/papers/volume15/srivastava14a/srivastava14a.pdf`.

- Zhang, Z., Ma, H., Fu, H., & Zha, C. (2016). *Scene-Free Multi-Class Weather Classification on Single Images.* IEEE Access, 8, 146,038–146,049. doi:10.1109: `https://web.cse.ohio-state.edu/~zhang.7804/Cheng_NC2016.pdf`.

9

Transfer Learning

One of the most significant developments of the last decade in the **machine learning** (**ML**) space was the concept of **transfer learning**, and rightfully so. Transfer learning is the process of applying knowledge gained from solving a source task to a target task, which is a different but related task. This approach has proven not only effective in saving computational resources required to train a deep neural network but also in cases where the target dataset is limited in size. Transfer learning reuses learned features from a pre-trained model, enabling us to build better-performing models and attain convergence much faster. Because of its numerous benefits, transfer learning has become an area of extensive research, with several studies exploring the application of transfer learning across different domains, such as image classification, object detection, natural language processing, and speech recognition.

In this chapter, we will introduce the concept of transfer learning, examining how it works, and some best practices around the application of transfer learning in various use cases. We will apply the concept of transfer learning in a real-world application with the aid of well-known pre-trained models. We will see in action how to apply these pre-trained models as a feature extractor and also learn how to fine-tune them to achieve optimal results. By the end of this chapter, you will have a solid understanding of what transfer learning is and how to apply it effectively to build real-world image classifiers.

In this chapter, we will cover the following topics:

- An introduction to transfer learning
- Types of transfer learning
- Building a real-world image classifier with transfer learning

Technical requirements

We will use Google Colab to run the coding exercise, which requires `python >= 3.8.0`, along with the following packages, which can be installed using the `pip install` command:

- `tensorflow>=2.7.0`

- `os`

- `matplotlib >=3.4.0`

- `pathlib`

The code bundle for this book is available at the following GitHub link: `https://github.com/PacktPublishing/TensorFlow-Developer-Certificate`. Also, solutions to all exercises can be found in the GitHub repo itself.

Introduction to transfer learning

As humans, it is easy for us to transfer knowledge gained from one task or activity to another. For instance, if you have a good grasp of Python (the programming language, not the snake) and you decide to learn Rust, because of your background knowledge in Python, you will find it easier to learn Rust compared to someone who has never written a basic program in any programming language. This is because certain concepts, such as object-oriented programming, have similarities across different programming languages. Transfer learning follows the same principle.

Transfer learning is a technique in which we leverage a model pre-trained on *task A* to solve a different but related *task B*. For example, we use a neural network trained on one task and transfer the knowledge gained to multiple related tasks. In image classification, we often use deep learning models that have been trained on very large datasets, such as ImageNet, which is made up of more than 1,000,000 images across 1,000 categories. The knowledge gained by these pre-trained models can be applied to many different tasks, such as classifying different breeds of dogs in a photograph. Just like how we can learn Rust quicker because of our knowledge of Python, the same applies here – pre-trained models can leverage information gained from a source task and apply it to the target task, reducing both the training time and the need for a large amount of annotated data, which may not be available or difficult to collect for the target task.

Transfer learning is not limited to image classification tasks; it can also be applied to other deep learning tasks, such as natural language processing, speech recognition, and object detection. In *Chapter 11, NLP with TensorFlow*, we will apply transfer learning to text classification. There, we will see how pretrained models (which we will access from TensorFlow Hub), trained on large text corpora, can be fine-tuned for text classification.

In classic ML, as illustrated in *Figure 9.1(a)*, we train the model from scratch for each task, as we have done so far in this book. This approach is resource- and data-intensive.

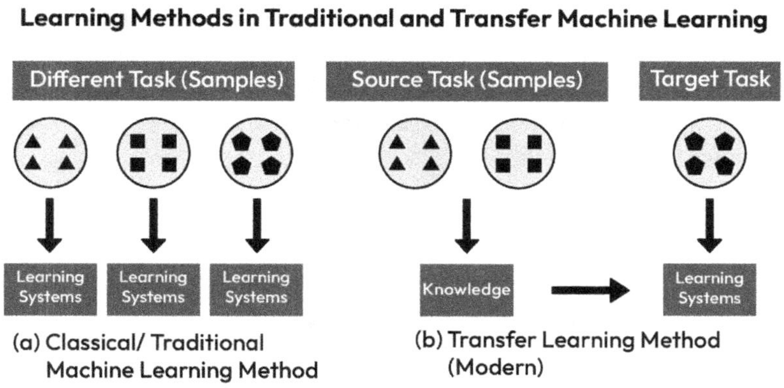

Learning Methods in Traditional and Transfer Machine Learning

Figure 9.1 – Traditional ML versus transfer learning

However, researchers discovered it was possible to learn visual features so that a model learns low-level features from a massive dataset, such as ImageNet, and applies this to a new, related task, as illustrated in *Figure 9.1(b)* – for example, in the classification of our weather dataset, which we used in *Chapter 8, Handling Overfitting*. By applying transfer learning, we can take advantage of the knowledge gained by a model during its training on a large dataset and adapt it to solve different but related tasks effectively. This approach proved useful, as it not only saves training time and resources but has also learned to improve performance, even in scenarios where a limited amount of data is available for the target task.

Types of transfer learning

There are two main ways we can apply transfer learning in CNNs. First, we can use the pre-trained model as a feature extractor. Here, we freeze the weights of the convolutional layers to preserve the knowledge gained from the source task and add a classifier, which is trained for classification of the second task. This works because the convolutional layers are reusable, since they only learned the low-level features such as edges, corners, and textures, which are generic and applicable in different images, as shown in *Figure 9.2*, while the fully connected layers are added to learn high-level details, which are used to classify different objects in a photograph.

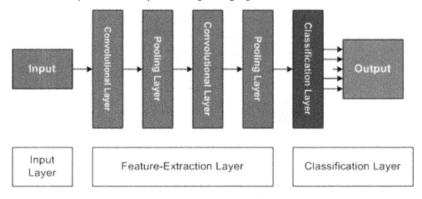

Figure 9.2 – Transfer learning as a feature extractor

The second method of applying transfer learning is to unfreeze some layers of the pre-trained model and add a classifier model to identify the high-level features, as shown in *Figure 9.3*. Here, we train both the unfrozen layers and the new classifier together. The pre-trained model is applied as the starting point of the new task, and the weight of the unfrozen layers is fine-tuned along with the classification layer to adapt the model to the new task.

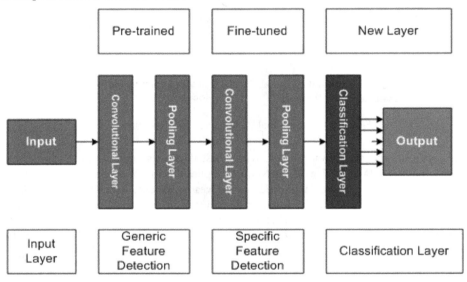

Figure 9.3 – Transfer learning as a fine-tuned model

Pre-trained models are deep networks that have been trained on large datasets. By leveraging the knowledge and weights these models have already acquired, we can use them as a feature extractor or fine-tune them for our use case, with a smaller dataset and less training time. Transfer learning provides ML practitioners with access to state-of-the-art models, which can be quickly and easily accessed in TensorFlow using an API. This means we don't always have to train our model from scratch, saving time and computational resources, as fine-tuning a model is faster than training it from the beginning.

We can apply pre-trained models to relevant use cases, potentially leading to higher accuracy and faster convergence. However, if the source and target domains are unrelated, transfer learning may not only fail but also harm the performance of the target task, due to irrelevant learned features, a situation known as negative transfer. Let's apply transfer learning to a real-world image classification task. We will explore some of the top-performing pretrained models, such as VGG, Inception, MobileNetV2, and EfficientNet. These models have been pretrained for image classification tasks. Let's see how they will fare on the given task.

Building a real-world image classifier with Transfer learning

In this case study, your company secured a medical project, and you are assigned the responsibility to build a pneumonia classifier for GETWELLAI. You have been provided with over 5,000 X-ray JPEG images, made up of two categories (pneumonia and normal). The dataset was annotated by expert physicians and low-quality images have been removed. Let's see how we can tackle this problem using the two types of transfer learning techniques we have discussed so far.

Loading the data

Perform the following steps to load the data:

1. As usual, we start by loading the necessary libraries that we will need for our project:

    ```
    #Import necessary libraries
    import os
    import pathlib
    import matplotlib.pyplot as plt
    import matplotlib.image as mpimg
    import random
    import numpy as np
    from PIL import Image
    import pandas as pd
    import tensorflow as tf
    from tensorflow import keras
    from tensorflow.keras.preprocessing.image import
    ImageDataGenerator
    from tensorflow.keras.callbacks import EarlyStopping
    from tensorflow.keras import regularizer
    ```

2. Next, let's load the X-ray dataset. To do this, we will use the `wget` command to download the file from the specified URL:

    ```
    !wget https://storage.googleapis.com/x_ray_dataset/dataset.zip
    ```

3. The downloaded file is saved in the current working directory of our Colab instance as a ZIP file, which contains a dataset of the X-ray images.

4. Next, we will extract the contents of the `zip` folder by running the following code:

    ```
    !unzip dataset.zip
    ```

    When we run the code, we extract a `dataset` folder that holds `test`, `val`, and `train` sub-directories, with each sub-directory holding data for both normal and pneumonia X-ray images, as shown in *Figure 9.4*:

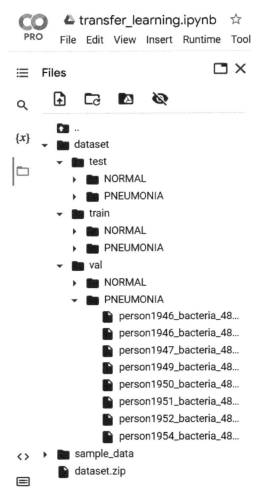

Figure 9.4 – A snapshot of the current working directory that holds the extracted ZIP file

5. We will use the following code block to extract the sub-directories and the number of images in them. We also saw this code block in *Chapter 8, Handling Overfitting*:

```
root_dir = "/content/dataset"

for dirpath, dirnames, filenames in os.walk(root_dir):
    print(f"Directory: {dirpath}")
    print(f"Number of images: {len(filenames)}")
    print()
```

It gives us a snapshot of the data in each folder and a sense of the data distribution.

6. Next, we will use the `view_random_images` function to display some random images and their shapes from the `train` directory:

```
view_random_images(
    target_dir="/content/dataset/train",num_images=4)
```

When we run the code, we will get a result similar to *Figure 9.5*.

```
Shape of image 1: (1978, 1979)
Shape of image 2: (848, 1296)
Shape of image 3: (1104, 1608)
Shape of image 4: (2363, 2288)
```

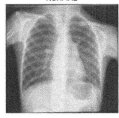 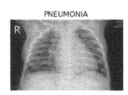 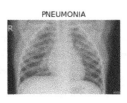 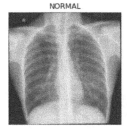

Figure 9.5 – Random images displayed from the training samples of the X-ray dataset

7. We will create an instance of the `ImageDataGenerator` class for our training and validation data. We will add the `rescale` parameter to rescale our images and ensure that all the pixel values are within the range of 0 to 1. We do this to improve stability and enhance convergence during the training process. The resulting `train_datagen` and `valid_datagen` objects are used to generate batches of training and validation data, respectively:

```
train_datagen = ImageDataGenerator(rescale=1./255)
valid_datagen = ImageDataGenerator(rescale=1./255)
```

8. Next, we set up the `train`, `validation`, and `test` directories.

```
# Set up the train and test directories
train_dir = "/content/dataset/train/"
val_dir = "/content/dataset/val"
test_dir = "/content/dataset/test"
```

9. We used the `flow_from_directory()` method to load images from the training directory. The `target_size` argument is used to resize all the images to 224 x 224 pixels. One key difference between this code and the one we used in *Chapter 8, Handling Overfitting* is that the class mode argument is set to `binary` because we are dealing with a binary classification problem (i.e., normal and pneumonia):

```
train_data=train_datagen.flow_from_directory(
    train_dir,target_size=(224,224),
# convert all images to be 224 x 224
```

```
                class_mode="binary")

        valid_data=valid_datagen.flow_from_directory(val_dir,
            target_size=(224,224),
            class_mode="binary",
            shuffle=False)
        test_data=valid_datagen.flow_from_directory(test_dir,
            target_size=(224,224),
            class_mode="binary",
            shuffle=False)
```

The valid_data and test_data generators are quite similar to the train_data generator, as they both have their target size set to 224 x 224 as well; the key difference is they have set shuffle to false, which means the images will not be shuffled. If we set this to true, the images get shuffled.

Modeling

We will start by using the same model we applied in *Chapter 8, Handling Overfitting*. To avoid redundancy, let's focus on the fully connected layer, where we have one neuron in the output layer, as this is a binary classification task. We will compare the result with using transfer learning:

```
    tf.keras.layers.Flatten(),
    tf.keras.layers.Dense(1050, activation="relu"),
    tf.keras.layers.Dense(1, activation="sigmoid")
])

# Compile the model
model_1.compile(loss="binary_crossentropy",
    optimizer=tf.keras.optimizers.Adam(),
    metrics=["accuracy"])

#Fit the model
# Add an early stopping callback
callbacks = [tf.keras.callbacks.EarlyStopping(
    monitor="val_accuracy", patience=3,
    restore_best_weights=True)]
history_1 = model_1.fit(train_data,epochs=20,
    validation_data=valid_data,
    callbacks=[callbacks]
```

In this case, we have one neuron in the output layer, and we changed the activation function to a sigmoid function, since we are building a binary classifier. In the compile step, we also change the loss function to binary cross entropy; everything else remains the same. Then, we fit the model.

The training ends on the 7th epoch, as the validation loss fails to drop further:

```
Epoch 4/20
163/163 [==============================] - 53s 324ms/step - loss:
0.0632 - accuracy: 0.9774 - val_loss: 0.0803 - val_accuracy: 1.0000
Epoch 5/20
163/163 [==============================] - 53s 324ms/step - loss:
0.0556 - accuracy: 0.9797 - val_loss: 0.0501 - val_accuracy: 1.0000
Epoch 6/20
163/163 [==============================] - 53s 323ms/step - loss:
0.0412 - accuracy: 0.9854 - val_loss: 0.1392 - val_accuracy: 0.8750
Epoch 7/20
163/163 [==============================] - 54s 334ms/step - loss:
0.0314 - accuracy: 0.9875 - val_loss: 0.2450 - val_accuracy: 0.8750
```

On the fifth epoch, the model reached a validation accuracy of 100 percent, which looks promising. Let's evaluate the model:

```
model_1.evaluate(test_data)
```

When we evaluate the model on test data, we recorded only an accuracy of 0.7580. This points to signs of overfitting. Of course, we can try a combination of the ideas we learned in *Chapter 8, Handling Overfitting* to improve the model's performance and you are encouraged to do so. However, let's learn how to use pre-trained models and see whether we can transfer the knowledge gained by these models to our use case and, if possible, get better results. Let's do that next.

Modeling with transfer learning

In this section, we will utilize three widely used pre-trained CNNs for image classification – VGG16, InceptionV3, and MobileNet. We will demonstrate the application of transfer learning as a feature extractor using these models, followed by adding a fully connected layer for label classification. We will also learn how to fine-tune a pre-trained model by unfreezing some of its layers. Before we can use these models, we need to import them. We can do this using a single line of code:

```
from tensorflow.keras.applications import InceptionV3,
    MobileNet, VGG16, ResNet50
```

Now that we have our models and we are set to go, let's begin with VGG16.

VGG16

VGG16 is a CNN architecture developed by the Visual Geometry Group at the University of Oxford. It was trained on the ImageNet dataset. The VGG16 architecture secured second place in the image classification category in the ImageNet Challenge 2014 submission. VGG16 is made up of 13 (a 3 x 3 filter) convolutional layers, 5 (2x2) max-pooling layers, and 3 fully connected layers, as illustrated in *Figure 9.6.*

This gives us 16 layers with learnable parameters; recall that max-pooling layers are for dimensionality reduction and they have no weight. This one takes an input tensor of the 224 x 224 RGB image.

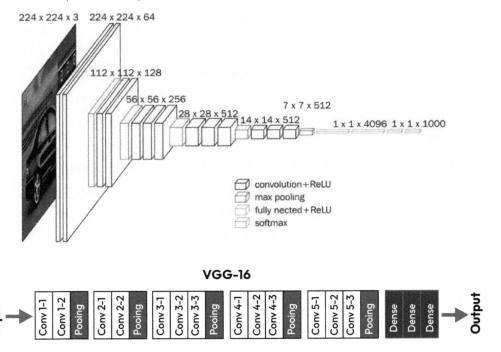

Figure 9.6 – The VGG16 model's architecture (Source: https://medium.com/analytics-vidhya/car-brand-classification-using-vgg16-transfer-learning-f219a0f09765)

Let's begin by loading VGG16 from Keras. We want to load the model and use pre-trained weights from the ImageNet dataset. To do this, we set the `weights` parameter to `imagenet`; we also set the `include_top` parameter to `false`. This is done because we want to use the model as a feature extractor. This way, we can add our own custom-made, fully connected layers for classification. We set the input size to (224,224,3) as this is the input image size that VGG16 expects:

```
# Instantiate the VGG16 model
vgg16 = VGG16(weights='imagenet', include_top=False,
    input_shape=(224, 224, 3))
```

The next step enables us to freeze the weights of the model because we want to use VGG 16 as a feature extractor. When we freeze all the layers, this makes them untrainable, which means their weights will not be updated during training:

```
# Freeze all layers in the VGG16 model
for layer in vgg16.layers:
    layer.trainable = False
```

The next code block creates a new sequential model that uses VGG as its top layer, after which we add a fully connected layer made up of a dense layer with 1,024 neurons, a dropout layer, and an output layer with one neuron, and then we set the activation to sigmoid for binary classification:

```
# Create a new model on top of VGG16
model_4 = tf.keras.models.Sequential()
model_4.add(vgg16)
model_4.add(tf.keras.layers.Flatten())
model_4.add(tf.keras.layers.Dense(1024, activation='relu'))
model_4.add(tf.keras.layers.Dropout(0.5))
model_4.add(tf.keras.layers.Dense(1, activation='sigmoid'))
```

We compile and fit our model to the data:

```
# Compile the model
model_4.compile(optimizer='adam',
    loss='binary_crossentropy', metrics=['accuracy'])

# Fit the model
callbacks = [tf.keras.callbacks.EarlyStopping(
    monitor='val_accuracy', patience=3,
    restore_best_weights=True)]
history_4 = model_4.fit(train_data,
    epochs=20,
    validation_data=valid_data,
    callbacks=[callbacks]
    )
```

In four epochs, our model stops training. It reaches a training accuracy of 0.9810 but on the validation set, we get an accuracy of 0.875:

```
Epoch 1/20
163/163 [==============================] - 63s 360ms/step - loss:
0.2737 - accuracy: 0.9375 - val_loss: 0.2021 - val_accuracy: 0.8750
Epoch 2/20
163/163 [==============================] - 57s 347ms/step - loss:
0.0818 - accuracy: 0.9699 - val_loss: 0.4443 - val_accuracy: 0.8750
Epoch 3/20
163/163 [==============================] - 56s 346ms/step - loss:
0.0595 - accuracy: 0.9774 - val_loss: 0.1896 - val_accuracy: 0.8750
Epoch 4/20
163/163 [==============================] - 58s 354ms/step - loss:
0.0556 - accuracy: 0.9810 - val_loss: 0.4209 - val_accuracy: 0.8750
```

When we evaluate the model, we reach an accuracy of 84.29. Now, let's use another pre-trained model as a feature extractor.

MobileNet

MobileNet is a lightweight CNN model developed by engineers at Google. The model is light and efficient, making it a choice model to develop mobile and embedded vision apps. Like VGG16, MobileNet was also trained on the ImageNet dataset, and it was able to achieve state-of-the-art results. MobileNet has a streamlined architecture that makes use of depth-wise separable convolutions. The underlining idea is to reduce the number of parameters required during training while maintaining accuracy.

To apply MobileNet as a feature extractor, the steps are similar to what we just did with VGG16; hence, let's look at the code block. We will load the model, freeze the layers, and add a fully connected layer as before:

```
# Instantiate the MobileNet model
mobilenet = MobileNet(weights='imagenet',
    include_top=False, input_shape=(224, 224, 3))

# Freeze all layers in the MobileNet model
for layer in mobilenet.layers:
    layer.trainable = False

# Create a new model on top of MobileNet
model_10 = tf.keras.models.Sequential()
model_10.add(mobilenet)
model_10.add(tf.keras.layers.Flatten())
model_10.add(tf.keras.layers.Dense(1024,activation='relu'))
model_10.add(tf.keras.layers.Dropout(0.5))
model_10.add(tf.keras.layers.Dense(1,activation='sigmoid'))
```

Next, we compile and fit the model:

```
# Compile the model
model_10.compile(optimizer='adam',
    loss='binary_crossentropy', metrics=['accuracy'])

# Fit the model
callbacks = [tf.keras.callbacks.EarlyStopping(
    monitor='val_accuracy', patience=3,
    restore_best_weights=True)]
history_10 = model_10.fit(train_data,
    epochs=20,
    validation_data=valid_data,
    callbacks=[callbacks])
```

In just four epochs, the model reaches a validation accuracy of 87.50%:

```
Epoch 1/20
163/163 [==============================] - 55s 321ms/step - loss:
3.1179 - accuracy: 0.9402 - val_loss: 1.8479 - val_accuracy: 0.8750
Epoch 2/20
163/163 [==============================] - 51s 313ms/step - loss:
0.3896 - accuracy: 0.9737 - val_loss: 1.1031 - val_accuracy: 0.8750
Epoch 3/20
163/163 [==============================] - 52s 320ms/step - loss:
0.0795 - accuracy: 0.9896 - val_loss: 0.8590 - val_accuracy: 0.8750
Epoch 4/20
163/163 [==============================] - 52s 318ms/step - loss:
0.0764 - accuracy: 0.9877 - val_loss: 1.1536 - val_accuracy: 0.8750
```

Next, let's try fine-tuning a pre-trained model hands-on.

Transfer learning as a fine-tuned model

InceptionV3 is another CNN architecture developed by Google. It combines 1x1 and 3x3 filters to capture different aspects of an image. Let's unfreeze some layers of this pre-trained model so that we can train both the layers we unfroze and the fully connected layer.

First, we will load the InceptionV3 model. We set include_top=False to remove the classification layer of InceptionV3 and use weights from ImageNet. We unfreeze the last 50 layers by setting trainable to true for these layers. This enables us to train these layers on the X-ray dataset:

```
# Load the InceptionV3 model
inception = InceptionV3(weights='imagenet',
    include_top=False, input_shape=(224, 224, 3))

# Unfreeze the last 50 layers of the InceptionV3 model
for layer in inception.layers[-50:]:
    layer.trainable = True
```

> **Note:**
> Unfreezing and fine-tuning too many layers on a small dataset is a bad strategy, as this can lead to overfitting.

We will create, fit, and compile the model, as we have done so far, and the new model reaches a validation accuracy of 100 percent on the fifth epoch:

```
Epoch 5/10
163/163 [==============================] - 120s 736ms/step - loss:
0.1168 - accuracy: 0.9584 - val_loss: 0.1150 - val_accuracy: 1.0000
```

```
Epoch 6/10
163/163 [==============================] - 117s 716ms/step - loss:
0.1098 - accuracy: 0.9624 - val_loss: 0.2713 - val_accuracy: 0.8125
Epoch 7/10
163/163 [==============================] - 123s 754ms/step - loss:
0.1011 - accuracy: 0.9613 - val_loss: 0.2765 - val_accuracy: 0.7500
Epoch 8/10
163/163 [==============================] - 120s 733ms/step - loss:
0.0913 - accuracy: 0.9668 - val_loss: 0.2711 - val_accuracy: 0.8125
```

Next, let's evaluate the models using our `evaluate_models` helper function:

	Model	Loss	Accuracy
0	model 10	36.73	87.82
1	model 4	31.57	87.66
2	model 8	35.41	87.66
3	model 9	56.47	84.78
4	model 6	78.13	84.29
5	model 3	55.84	81.89
6	model 2	48.09	81.25
7	model 1	118.34	75.96
8	model 5	54.05	68.91
9	model 7	67.82	61.06

Figure 9.7 – An evaluation result from our experiments

From the results in *Figure 9.7*, MobileNet, VGG16, and InceptionV3 came out on top. We can see that these models performed much better than our baseline model (**model 1**). We also report the results of a few other models from our notebook. We can spot signs of overfitting; hence, you can combine some of the ideas we discussed in *Chapter 8, Handling Overfitting* to improve your result.

Summary

Transfer learning has gained traction in the deep learning community, due to its improved performance, speed, and accuracy in building deep learning models. We discussed the rationale behind transfer learning and explored transfer learning as a feature extractor and a fine-tuned model. We built a couple of solutions using the top-performing pre-trained models and saw how they outperformed our baseline model when applied to the X-ray dataset.

By now, you should have gained a solid understanding of transfer learning and its applications. Equipped with this knowledge, you should be able to apply transfer learning as either a feature extractor or a fine-tuned model when building real-world deep learning solutions for a wide range of tasks.

With this, we have come to the end of this chapter and this section of the book. In the next chapter, we will discuss **natural language processing** (**NLP**), where we will build exciting NLP applications using TensorFlow.

Questions

Let's test what we learned in this chapter:

1. Using the test notebook, load the cat and dog dataset.
2. Preprocess the image data using the image data generator.
3. Use a VGG16 model as a feature extractor and build a new CNN model.
4. Unfreeze 40 layers of an InceptionV3 model and build a new CNN model.
5. Evaluate both the VGG16 and InceptionV3 models.

Further reading

To learn more, you can check out the following resources:

- Kapoor, A., Gulli, A. and Pal, S. (2020) *Deep Learning with TensorFlow and Keras Third Edition: Build and deploy supervised, unsupervised, deep, and reinforcement learning models.* Packt Publishing Ltd.

- *Adapting Deep Convolutional Neural Networks for Transfer Learning: A Comparative Study* by C. M. B. Al-Rfou, G. Alain, and Y. Bengio, published in arXiv preprint arXiv:1511.

- *Very Deep Convolutional Networks for Large-Scale Image Recognition* by K. Simonyan and A. Zisserman, published in arXiv preprint arXiv:1409.1556 in 2014.

- *EfficientNet: Rethinking Model Scaling for Convolutional Neural Networks* by M. Tan and Q. Le, published in *International Conference on Machine Learning* in 2019.

- *MobileNetV2: Inverted Residuals and Linear Bottlenecks* by M. Sandler, A. Howard, M. Zhu, A. Zhmoginov, and L. Chen, published in arXiv preprint arXiv:1801.04381 in 2018.

- *DeCAF: A Deep Convolutional Activation Feature for Generic Visual Recognition* by Donahue, J., Jia, Y., Vinyals, O., Hoffman, J., Zhang, N., Tzeng, E., & Darrell, T. (2014).

- *Harnessing the power of transfer learning for medical image classification* by Ryan Burke, *Towards Data Science.* https://towardsdatascience.com/harnessing-the-power-of-transfer-learning-for-medical-image-classification-fd772054fdc7

Part 3 –
Natural Language Processing
with TensorFlow

In this part, you will learn to build **natural language processing** (**NLP**) applications with TensorFlow. You will understand how to perform text processing and build models for text classification. In this part, you will also learn to generate text using LSTMs.

This section comprises the following chapters:

- *Chapter 10, Introduction to Natural Language Processing*
- *Chapter 11, NLP with TensorFlow*

10
Introduction to Natural Language Processing

Today, we are faced with a staggering amount of text data coming at us from all directions, from social media platforms to email communications, and from text messages to online reviews. This exponential growth in text data has led to a rapid growth in the development of text-based applications powered by advanced deep learning techniques, used to unlock insights from text data. We find ourselves in the dawn of a transformative era, powered by tech giants such as Google and Microsoft and revolutionary start-ups such as OpenAI and Anthropic. These visionaries are leading the charge in building powerful solutions capable of solving a myriad of text-based challenges, such as summarizing large volumes of documents, extracting sentiments from online platforms, and generating text for blog posts – the list of uses is endless.

Real-world text data can be messy; it could be riddled with unwanted information such as punctuation marks, special characters, and common words that may not contribute significantly to the text's meaning. Hence, we will kick off this chapter by looking at some basic text preprocessing steps to help transform text data into a more digestible form in preparation for modeling. Again, you may wonder, how do these machines learn to understand text? How do they make sense of words and sentences, or even grasp their semantic meaning or the context in which words are used? In this chapter, we will journey through the fundamentals of **natural language processing** (**NLP**). We will explore concepts such as tokenization, which deals with how we segment text into individual words or terms (tokens). We will also explore the concept of word embeddings – here, we will see how they enable models to capture the meaning, context, and relationship between words.

Then, we will put together all we have learned in this chapter to build a sentiment analysis model, using the Yelp Polarity dataset to classify customer reviews. As an interactive activity, we will examine how to visualize word embeddings in TensorFlow; this can be useful in gaining a snapshot of how our model understands and represents different words. We will also explore various techniques to improve the performance of our sentiment analysis classifier. By the end of this chapter, you will have a good foundational understanding of how to preprocess and model text data, as well as the skills required to tackle real-world NLP problems.

In this chapter, we will cover the following topics:

- Text preprocessing

- Building a sentiment classifier

- Embedding visualization

- Model improvement

Text preprocessing

NLP is an exciting and evolving field that lies at the intersection of computer science and linguistics. It empowers computers with the ability to understand, analyze, interpret, and generate text data. However, working with text data presents a unique set of challenges, one that differs from the tabular and image data we worked with in the earlier sections of this book. *Figure 10.1* gives us a high-level overview of some of the inherent challenges that text data presents. Let's drill into them and see what and how they present issues to us when building deep learning models with text data.

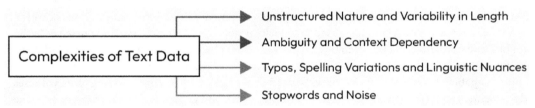

Figure 10.1 – The challenges presented by text data

Text data in its natural form is unstructured, and this is just the beginning of the uniqueness of this interesting type of data we will work with in this chapter. Let's illustrate some of the issues by looking at these two sentences – "*The house next to ours is beautiful*" versus "*Our neighbor's house is one everyone in this area admires.*" Both phrases have a similar sentiment, yet they have different structures and varying lengths. To humans, this lack of structure and varying length is not a challenge, but when we work with deep learning models, this could pose a challenge. To address these challenges, we can consider ideas such as tokenization, which refers to the splitting of text data into smaller units called tokens. These tokens could be used to represent words, sub-words, or individual characters. To handle the varying length of text data in preparation for modeling, we will reuse an old trick that we applied when working with image data with CNNs – padding. By padding the sentences, we can ensure that our data (such as sentences or paragraphs) is of the same length. This uniformity makes our data more digestible for our models.

Again, we may come across words with multiple meanings and decoding the meaning of these words largely depends on the context in which they are used. For example, if we have a sentence reading "*I will be at the bank*," without additional context, it is difficult to tell whether *bank* refers to a financial bank or a riverbank. Words such as this add an extra layer of complexity when modeling text data. To handle this issue, we need to apply techniques that capture the essence of words and their surrounding words. A good example of such a technique is word embeddings. Word embeddings are powerful vector representations that can be used to capture the semantic meaning of words, by enabling words with a similar meaning or context to have similar representations.

Other issues we could face when working with text data are typos, spelling variations, and noise. To tackle these issues, we can use noise-filtering techniques to filter out URLs, special characters, and other irrelevant entities when collecting data online. Let's say we have a sample sentence – "*Max loves to play chess at the London country club, and he is the best golfer on our street*." When we examine this sentence, we see that it contains common words such as *and*, *is*, and *the*. Although these words are needed for linguistic coherence, in some instances, they may not add semantic value. If this is the case, then we may want to remove these words to reduce the dimensionality of our data.

Now that we've covered some foundational ideas around the challenges of text data, let's see how to preprocess text data by looking at how we can extract and clean text data on machine learning from Wikipedia. Here, we will see how to apply TensorFlow to perform techniques such as tokenization, padding, and using word embeddings to extract meaning from text. To access the sample data, use this link: `https://en.wikipedia.org/wiki/Machine_learning`. Let's begin:

1. We will start by importing the necessary libraries:

```
import requests
from bs4 import BeautifulSoup
import re
from tensorflow.keras.preprocessing.text import Tokenizer
```

We use these libraries to effectively fetch, preprocess, and tokenize web data, preparing it for modeling with neural networks. When we want to access data from the internet, the `requests` library can prove to be a useful tool, enabling us to streamline the process of retrieving information from web pages by making requests to web servers and fetching web data. The collected data is often in the HTML format, which isn't in the best shape for us to feed into our models. This is where `BeautifulSoup` (an intuitive tool for parsing HTML and XML) comes into the picture, enabling us to easily navigate and access the content we need. To perform string manipulation, text cleaning, or extracting patterns, we can use the `re` module. We also import the `Tokenizer` class from TensorFlow's Keras API, which enables us to perform tokenization, thus converting our data into a model-friendly format.

2. Then, we assign our variable to the web page we want to scrape; in this case, we are interested in scraping data from the Wikipedia page on machine learning:

```
# Define the URL of the page
url = "https://en.wikipedia.org/wiki/Machine_learning"

# Send a GET request to the webpage
response = requests.get(url)
```

We use the GET method to retrieve data from the web server. The web server replies with the status code that tells us whether the request was a success or failure. It also returns other metadata along with the HTML content of the web page – in our case, the Wikipedia page. We save the server's response to the GET request in the response object.

3. We use the BeautifulSoup class to do the heavy lifting of parsing the HTML content, which we access by using response.content:

```
# Parse the HTML content of the page with BeautifulSoup
soup = BeautifulSoup(response.content, 'html.parser')
```

Here, we convert the raw HTML contents contained in response.content into a digestible format by specifying html.parser and storing the result in the soup variable.

4. Now, let's extract all the text contents in a paragraph:

```
# Extract the text from all paragraph tags on the page
passage = " ".join([
    p.text for p in soup.find_all('p')])
```

We use soup.find_all('p') to extract all the paragraphs stored in the soup variable. Then, we apply the join method to combine them into a body of text in which each paragraph is repeated by a space, and then we store this text in the passage variable.

5. The next step is the removal of stopwords from our data. Stopwords are common words that may not contain useful information in a certain use case. Hence, we may want to remove them to help reduce the dimensionality of our data, especially for tasks where these high-frequency words offer little importance, such as information retrieval or document clustering. Here, it may be wise to remove stopwords to enable faster convergence and produce better categorization. Examples of stopwords are words such as "and," "the," "in," and "is":

```
# Define a simple list of stopwords
stopwords = ["i", "me", "my", "myself", "we", "our",
    "ours", "ourselves", "you", "your",
    "yours", "yourself", "yourselves", "he",
    "him", "his", "himself", "she", "her",
```

```
"hers", "herself", "it", "its", "itself",
"they", "them", "their", "theirs",
"themselves", "what", "which", "who",
"whom", "this", "that", "these", "those",
"am", "is", "are", "was", "were", "be",
"been", "being", "have", "has", "had",
"having", "do", "does", "did", "doing",
"a", "an", "the", "and", "but", "if",
"or", "because", "as", "until", "while",
"of", "at", "by", "for", "with", "about",
"against", "between", "into", "through",
"during", "before", "after", "above",
"below", "to", "from", "up", "down",
"in", "out", "on", "off", "over",
"under", "again", "further", "then",
"once", "here", "there", "when", "where",
"why", "how", "all", "any", "both",
"each", "few", "more", "most", "other",
"some", "such", "no", "nor", "not",
"only", "own", "same", "so", "than",
"too", "very", "s", "t", "can", "will",
"just", "don", "should", "now"]
```

We have defined a list of stopwords. This way, we have the flexibility of adding words of our choice to this list. This approach can be useful when working on domain-specific projects in which you may want to extend your stopword list.

> **Note**
>
> This step might not always be beneficial. In some NLP tasks, stopwords might contain useful information. For example, in text generation or machine translation, a model needs to generate/translate stopwords to produce coherent sentences.

6. Let's convert the entire passage into lowercase. We do this to ensure words with the same semantic meaning are not represented differently – for example, "DOG" and "dog." By ensuring all our data is in lowercase, we introduce uniformity to our dataset, removing the possibility of duplicate representation of the same word. To convert our text to lowercase, we use the following code:

```
passage = passage.lower()
```

When we run the code, it converts all our text data to lowercase.

> **Note**
>
> Converting a body of text to lowercase isn't always the best solution. In fact, in some use cases, such as sentiment analysis, converting to lowercase may lead to information loss because capital letters are usually used to express strong emotions.

7. Next, let's remove the HTML tags, special characters, and stopwords from the passage we gathered from Wikipedia. To do this, we'll use the following code:

```
# Remove HTML tags using regex
passage = re.sub(r'<[^>]+>', '', passage)
# Remove unwanted special characters
passage = re.sub('[^a-zA-Z\s]', '', passage)

# Remove stopwords
passage = ' '.join(word for word in passage.split() if word not
in stopwords)

# Print the cleaned passage
print(passage[:500])  # print only first 500 characters for
brevity
```

In the first line of code, we remove the HTML tags, after which we remove unwanted special characters in our data. Then, we pass the passage through a stopword filter to check and remove words that are in the stopword list, after which we combine the remaining words into a passage, separated by spaces between them. We print the first 500 characters to get an idea of what our processed text looks like.

8. Let's print the first 500 characters and compare that with the web page shown in *Figure 10.2*:

```
machine learning ml field devoted understanding building
methods let machines learn methods leverage data improve
computer performance set tasks machine learning algorithms
build model based sample data known training data order make
predictions decisions without explicitly programmed machine
learning algorithms used wide variety applications medicine
email filtering speech recognition agriculture computer vision
difficult unfeasible develop conventional algorithms perform
needed tasks subset ma
```

When we compare the output with the web page, we can see that our text is all in lowercase and all the stopwords, special characters, and HTML tags have been removed.

Machine learning

文A 73 languages ∨

Article   Talk

Read   Edit   View history   Tools ∨

From Wikipedia, the free encyclopedia

For the journal, see Machine Learning *(journal).*

"Statistical learning" redirects here. For statistical learning in linguistics, see statistical learning in language acquisition.

Machine learning (ML) is a field devoted to understanding and building methods that let machines "learn" – that is, methods that leverage data to improve computer performance on some set of tasks.[1]

Machine learning algorithms build a model based on sample data, known as training data, in order to make predictions or decisions without being explicitly programmed to do so.[2] Machine learning algorithms are used in a wide variety of applications, such as in medicine, email filtering, speech recognition, agriculture, and computer vision, where it is difficult or unfeasible to develop conventional algorithms to perform the needed tasks.[3][4]

A subset of machine learning is closely related to computational statistics, which focuses on making predictions using computers, but not all machine learning is statistical learning. The study of mathematical optimization delivers methods, theory and application domains to the field of machine learning. Data mining is a related field of study, focusing on exploratory data analysis through unsupervised learning.[6][7]

Some implementations of machine learning use data and neural networks in a way that mimics the working of a biological brain.[8][9]

In its application across business problems, machine learning is also referred to as predictive analytics.

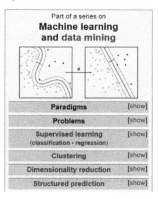

Part of a series on
**Machine learning
and data mining**

Paradigms	[show]
Problems	[show]
Supervised learning (classification · regression)	[show]
Clustering	[show]
Dimensionality reduction	[show]
Structured prediction	[show]

Figure 10.2 – A screenshot of the Wikipedia page on machine learning

Here, we explored some simple steps in preparing text data for modeling with neural networks. Now, let's extend our learning by exploring tokenization.

Tokenization

We have looked at some important ideas about how to preprocess real-world text data. Our next step is to map out a strategy to prepare our text model. To do this, let's begin by examining the concept of tokenization, which entails breaking sentences into smaller units called tokens. Tokenization can be implemented at character, sub-word, word, or even sentence level. It is common to see a lot of word-level tokenizers; however, the choice of tokenizer to use largely depends on the use case.

Let's see how we can apply tokenization to a sentence with a sample. Let's say we have this sentence – "*I like playing chess in my leisure time.*" Applying word-level tokenization will give us output such as `["i", "like", "playing", "chess", "in", "my", "leisure", "time"]`, while if we decide to use character level tokenization, we will have output such as `['I', ' ', 'l', 'i', 'k', 'e', ' ', 'p', 'l', 'a', 'y', 'i', 'n', 'g', ' ', 'c', 'h', 'e', 's', 's', ' ', 'i', 'n', ' ', 'm', 'y', ' ', 'l', 'e', 'i', 's', 'u', 'r', 'e', ' ', 't', 'i', 'm', 'e']`. In word-level tokenization, we split across each word, while in character-level tokenization, we split on each character. You can also see that wide spaces within the sentence are included when we use character-level tokenization. For subword and sentence-level tokenization, we split into subwords and sentences respectively. Now, let's use TensorFlow to implement word-level and character-level tokenization.

Word-level tokenization

Let's see how to perform word-level tokenization with TensorFlow:

1. We will begin by importing the `Tokenizer` class:

    ```
    from tensorflow.keras.preprocessing.text import Tokenizer
    ```

2. Then, we create a variable called text to hold our sample sentence (`"Machine learning is fascinating. It is a field full of challenges!"`). We create an instance of the `Tokenizer` class to handle the tokenization of our sample sentence:

    ```
    text = "Machine learning is fascinating. It is a field full of
    challenges!"
    # Define the tokenizer and fit it on the text
    tokenizer = Tokenizer()
    tokenizer.fit_on_texts([text])
    ```

    We can pass several parameters into the `tokenizer` class, depending on our use case. For instance, we can set the maximum number of words we want to keep by using `num_words`. Also, we may want to convert our entire text into lowercase; we can do this with the `tokenizer` class. However, if we don't specify these parameters, TensorFlow will apply the default parameters. Then, we use the `fit_on_text` method to fit the tokenizer on our text data. The `fit_on_text` method goes through the input text and creates a vocabulary made up of unique words. It also counts the number of occurrences of each word in our input text data.

3. To view the mapping of words to integer values, we use the `word_index` property of our `tokenizer` object:

    ```
    # Print out the word index to see how words are tokenized
    print(tokenizer.word_index)
    ```

    When we print out the result, we can see that `word_index` returns a dictionary of key-value pairs, where each key-value pair corresponds to a unique word and its assigned integer index in the tokenizer's vocabulary:

    ```
    {'is': 1, 'machine': 2, 'learning': 3, 'fascinating': 4, 'it':
    5, 'a': 6, 'field': 7, 'full': 8, 'of': 9, 'challenges': 10}
    ```

You can see that the exclamation mark in our sample sentence is gone and the word `'is'` is listed only once. Also, you can see that our indexing begins at 1 and not 0, because 0 is reserved as a special token, which we will encounter shortly. Now, let's also examine how to perform character-level tokenization.

Character-level tokenization

In character-level tokenization, we split the text on each character within our sample text. To do this with TensorFlow, we slightly modify the code we used for word-level tokenization:

```
# Define the tokenizer and fit it on the text
tokenizer = Tokenizer(char_level=True)
tokenizer.fit_on_texts([text])
# Print out the character index to see how characters are tokenized
print(tokenizer.word_index)
```

Here, we set the `char_level` argument to `True` when we create our `tokenizer` instance. When we do this, we can see that only unique characters in our text will be treated as separate tokens:

```
{' ': 1, 'i': 2, 'a': 3, 'n': 4, 'l': 5, 'e': 6, 's': 7,
 'f': 8, 'c': 9, 'g': 10, 'h': 11, 't': 12, 'm': 13,
 'r': 14, '.': 15, 'd': 16, 'u': 17, 'o': 18, '!': 19}
```

Note that every unique character is represented in this scenario, including whitespaces (`' '`) with a token value of 1, full stops (`'.'`) with a token value of 15, and exclamations (`'!'`) with a token value of 19.

Another type of tokenization we talked about is subwords. Subwords involve breaking down words into commonly occurring groups of characters – for example, "unhappiness" might be tokenized into ["un", "happiness"]. Once the text is tokenized, each token can be transformed into a numerical representation, using one of the encoding methods that we will discuss in this chapter. Now, let's look at another concept called sequencing.

Sequencing

The order in which words are used in a sentence is crucial to understanding the meaning they convey; sequencing is the process of converting sentences or a group of words or tokens into their numerical representations, preserving the sequential order of words when building NLP applications using neural networks. In TensorFlow, we can use the `texts_to_sequences` function to convert our tokenized text into a sequence of integers. From the output of our word-level tokenization step, we now know that our sample sentence (`"Machine learning is fascinating. It is a field full of challenges!"`) can be represented by a list of tokens:

```
# Convert the text to sequences
sequence = tokenizer.texts_to_sequences([text])
print(sequence)
```

By converting text into sequences, we translate human-readable text into a machine-readable format while preserving the order in which words occur. When we print the result, we get the following:

```
[[2, 3, 1, 4, 5, 1, 6, 7, 8, 9, 10]]
```

The output printed is the sequence of integers that represent the original text. In many real-world scenarios, we will have to handle sentences of varying lengths. While it is not a problem for humans to understand sentences irrespective of their length, neural networks require us to put our data in a defined type of input format. In the image classification section of this book, we used a fixed width and height when passing image data as input; with text data, we have to ensure that all our sentences are of the same length. To do this, let's return to a concept we discussed in *Chapter 7, Image Classification with Convolutional Neural Networks,* padding, and see how we can leverage it to resolve the issue of varying sentence lengths.

Padding

In *Chapter 7, Image Classification with Convolutional Neural Networks,* we introduced the concept of padding when we discussed CNNs. In the context of NLP, padding is the process of adding elements to a sequence to ensure that it attains a desired length. To do this in TensorFlow, we use the `pad_sequences` function from the `keras` preprocessing module. Let's use an example to explain the application of padding to text data:

1. Let's say we have the following four sentences:

```
sentences = [
    "I love reading books.",
    "The cat sat on the mat.",
    "It's a beautiful day outside!",
    "Have you done your homework?"
]
```

When we perform word-level tokenization and sequencing, the output will look like this:

```
[[2, 3, 4, 5], [1, 6, 7, 8, 1, 9],
    [10, 11, 12, 13, 14], [15, 16, 17, 18, 19]]
```

We can see that the length of our returned sequences varies with the second sentence longer than the other sentences. Let's resolve this issue using padding next.

2. We import `pad_sequences` from the TensorFlow Keras preprocessing module:

```
from tensorflow.keras.preprocessing.sequence import pad_
sequences
```

The `pad_sequences` function takes various parameters – here, we will discuss a few important ones, such as `sequences`, `maxlen`, `truncating`, and `padding`.

3. Let's start passing sequences as the only parameter and observe what the result looks like:

```
padded = pad_sequences(sequences)
print(padded)
```

When we use the `pad_sequence` function, it ensures all the sentences are the same length as our longest sentence. To achieve this, a special token (0) is used to pad the shorter sentences until they are of the same length as the longest sentence. The special token (0) does not carry any meaning, and models are built to ignore them during training and inference:

```
[[ 0  0  2  3  4  5]
 [ 1  6  7  8  1  9]
 [ 0 10 11 12 13 14]
 [ 0 15 16 17 18 19]]
```

From the output, we see that every other sentence has zeros added to it until its length is the same length as our longest sentence (sentence two), which has the longest sequence. Note that all the zeros are added at the beginning of each sentence. This scenario is known as **prepadding**. We add the zeros at the end of each sentence by using the `padding=post` parameter:

```
# post padding
padded_sequences = pad_sequences(sequences,
    padding='post')
print(padded_sequences)
```

When we print the result, we get the following:

```
[[ 2  3  4  5  0  0]
 [ 1  6  7  8  1  9]
 [10 11 12 13 14  0]
 [15 16 17 18 19  0]]
```

In this case, we can see that the zeros are added at the end of shorter sentences.

4. Another useful parameter is `maxlen`. It is used to specify the maximum length for all sequences we want to keep. In this case, any sequence greater than the specified `maxlen` will be truncated. To see how `maxlen` works, let's add another sentence to our list of sentences:

```
Sentences = [
    "I love reading books.",
    "The cat sat on the mat.",
    "It's a beautiful day outside!",
    "Have you done your homework?",
    "Machine Learning is a very interesting subject that enables
     you build amazing solutions beyond your imagination."
]
```

5. We take our new list of sentences and perform tokenization and sequencing on them. Then, we pad the numerical representations to ensure that our input data is of the same length, and to ensure that our special (0) tokens are added at the end of a sentence, we set `padding` to `post`. When we implement these steps, our output looks like this:

```
[[ 5  6  7  8  0  0  0  0  0  0  0  0  0  0  0]
 [ 1  9 10 11  1 12  0  0  0  0  0  0  0  0  0]
```

```
[13   2 14 15 16   0   0   0   0   0   0   0   0   0   0   0]
[17   3 18   4 19   0   0   0   0   0   0   0   0   0   0   0]
[20 21 22   2 23 24 25 26 27   3 28 29 30 31   4 32]]
```

From the output, we can see that one sentence is quite long, and the other sentences are largely made up of numerical representations with many zeros. In this case, it's clear that our longest sentence is an outlier, since all the other sentences are much smaller. This can skew our model's learning process and also increase the computation resource required to model our data, especially when we work with large datasets or limited computation resources. To fix this situation, we apply the maxlen parameter. It is important to use a good max_length value; otherwise, we could lose important information in our data due to truncation. It is a good idea to make the maximum length long enough to capture useful information without adding much noise.

6.   Let's see how to apply maxlen in our example. We start by setting the max_length variable to 10. This means it will take only a maximum of 10 tokens. We pass the maxlen parameter and print our padded sequence:

```
# Define the max length
max_length = 10

# Pad the sequences
padded = pad_sequences(sequences, padding='post',
    maxlen=max_length)

print(padded))
```

Our result produces a much shorter sequence:

```
[[ 5   6   7   8   0   0   0   0   0   0]
 [ 1   9 10 11   1 12   0   0   0   0]
 [13   2 14 15 16   0   0   0   0   0]
 [17   3 18   4 19   0   0   0   0   0]
 [25 26 27   3 28 29 30 31   4 32]]
```

Note that our longest sentence has been truncated at the beginning of the sequence. What if we want to truncate the sentence at the end? How do we achieve this? To do this, we introduce another parameter called truncating and set it to post:

```
# Pad the sequences
padded = pad_sequences(sequences, padding='post',
    truncating='post', maxlen=max_length)

print(padded)
```

Our result will look like this:

```
[[ 5  6  7  8  0  0  0  0  0  0]
 [ 1  9 10 11  1 12  0  0  0  0]
 [13  2 14 15 16  0  0  0  0  0]
 [17  3 18  4 19  0  0  0  0  0]
 [20 21 22  2 23 24 25 26 27  3]]
```

We now have all our sequences padded and truncating done at the end of the sentence. Now, what if we train our model to classify text using these five sentences, and we want to make a prediction on a new sentence ("*I love playing chess*")? Remember from our training sentences that our model will have tokens to represent "I" and "love." However, it has no way of knowing or representing "playing" and "Chess." This presents us with another problem. Let's look at how to solve this.

Out of vocabulary

So far, we have seen how to prepare our data, moving from a sequence of text data that makes up sentences to numerical representations to train our models. Now, let's say we build a text classification model, which we train using the five sentences in our sentence list. Of course, this is a hypothetical situation, which would hold true even when we train with a massive amount of text data, as we will eventually come across words that our model has not seen before in training, such as in this scenario with our sample test sentence ("*I love playing chess*"). This means we must prepare our model to handle words that are not present in our predefined vocabulary. To fix this issue, we use **out-of-vocabulary** (**OOV**) tokens, and this can be done by instantiating the `oov_token="<OOV>"` argument during the tokenization process:

```
# Define the tokenizer with an OOV token
tokenizer = Tokenizer(oov_token="<OOV>")
# Fit the tokenizer on the texts
tokenizer.fit_on_texts(sentences)
# Convert the texts to sequences
sequences = tokenizer.texts_to_sequences(sentences)
# Let's look at the word index
print(tokenizer.word_index)
```

After this, we fit the tokenizer on the training sentences, converting the sentences to sequences of integers, and then we print out the word index:

```
{'<OOV>': 1, 'the': 2, 'a': 3, 'you': 4, 'your': 5, 'i': 6, 'love':
7, 'reading': 8, 'books': 9, 'cat': 10, 'sat': 11, 'on': 12, 'mat':
13, "it's": 14, 'beautiful': 15, 'day': 16, 'outside': 17, 'have': 18,
'done': 19, 'homework': 20, 'machine': 21, 'learning': 22, 'is': 23,
'very': 24, 'interesting': 25, 'subject': 26, 'that': 27, 'enables':
28, 'build': 29, 'amazing': 30, 'solutions': 31, 'beyond': 32,
'imagination': 33}
```

Now, we can see that the "<OOV>" string is chosen to represent these OOV words and has a value of 1. This token will take care of any unknown words that the model comes across. Let's see this in action with our sample test sentence:

```
# Now let's convert a sentence with some OOV words
test_sentence = "I love playing chess"
test_sequence = tokenizer.texts_to_sequences(
    [test_sentence])
print(test_sequence)
```

We pass in our test sentence ("I love playing chess"), which contains words our model has not seen before, and then use the texts_to_sequences method to convert the test sentence into a sequence. Because we fit the tokenizer on the training sentences, it will replace each word in the test sentence with its corresponding numerical representation from the word index. However, the words "playing" and "chess," which were not present in our training sentences, will be replaced with the index of the special OOV token; hence, the print statement returns the following:

```
[[6, 7, 1, 1]]
```

Here, the token value of 1 is used for the words playing and chess. Using the OOV token is a common practice in NLP to handle words that are not present in the training data but may appear in the test or real-world data.

Now, we have our text data as a numerical representation. We have also preserved the sequence in which words occur; however, we need to find a way to capture the semantic meaning of the words and their relationships with each other. To do this, we use word embeddings. Let's discuss word embeddings next.

Word embeddings

A significant landmark in the field of NLP is the use of word embeddings. With word embeddings, we are able to solve many complex modern-day text-based problems. **Word embeddings** are a type of word representation that allows words with similar meanings to have a similar representation, with the ability to capture the context in which a word is used. Along with the context, word embedding is also able to capture the semantic and syntactic similarity between words and how a word relates to other words. This allows ML models to generalize better when using word embedding in comparison to instances where words are used as standalone input.

Strategies such as one-hot encoding prove to be inefficient, as it builds a sparse representation of words largely made up of zeros. This happens because the more words we have in our vocabulary, the greater the number of zeros we will have in our resulting vector when we apply one-hot encoding. Conversely, word embedding is a dense vector representation in a continuous space that can capture the meaning, context, and relationship between words using dense and low-dimensional vectors.

Let's examine the following sample sentences and see how word embedding works:

1. She enjoys reading books.

2. He likes reading newspapers.

3. They are eating grapes.

We start by tokenizing each sentence and apply sequencing to transform each sentence into a sequence of integers:

1. [1, 2, 3, 4]

2. [5, 6, 3, 7]

3. [8, 9, 10, 11]

Observe that with our returned sequence, we have successfully captured the order in which the words that make up each sentence occur. However, this approach fails to take into consideration the meaning of words or the relationship between words. For example, the words "enjoy" and "likes" both portray positive sentiments in sentence 1 and sentence 2, while both sentences have "reading" as their common action. When we design deep learning models, we want them to be aware that "books" and "newspapers" are more closely related and differ from words such as "grapes" and "eating," as shown in *Figure 10.3*.

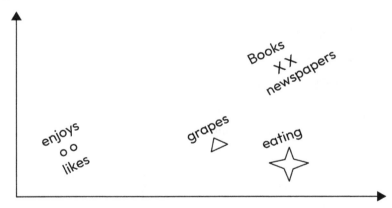

Figure 10.3 – Word embedding

Word embeddings empower our models to capture relationships between words, thus enhancing our ability to build better-performing models. We have explored some foundational ideas around text preprocessing and data preparation, taking our text data from words to numerical representations, capturing both sequencing and the underlying relationships between words used in language. Let's now put together everything we have learned and build a sentiment analysis model, using the Yelp Polarity dataset. We will start by training our own word embedding from scratch, after which we will apply pretrained word embeddings to our use case.

The Yelp Polarity dataset

In this experiment, we will work with the Yelp Polarity dataset. This dataset is made up of a training size of 560,000 reviews and 38,000 reviews for testing, with each entry consisting of a text-based review and a label (positive – 1 and negative – 0). The data was drawn from customer reviews of restaurants, hair salons, locksmiths, and so on. This dataset presents some real challenges – for example, the reviews are made up of text with varying lengths, from short reviews to very long reviews. Also, the data contains the use of slang and different dialects. The dataset is available at this link: https://www.tensorflow.org/datasets/catalog/yelp_polarity_reviews.

Let's start building our model:

1. We will begin by loading the required libraries:

    ```
    import tensorflow as tf
    import tensorflow_datasets as tfds
    from tensorflow.keras.preprocessing.text import Tokenizer
    from tensorflow.keras.preprocessing.sequence import pad_
    sequences
    from sklearn.model_selection import train_test_split
    import numpy as np
    import io
    ```

    We import the necessary libraries to load, split, preprocess, and visualize word embeddings, and model our dataset with TensorFlow for our sentiment analysis use case.

2. Then, we load the dataset:

    ```
    # Load the Yelp Polarity Reviews dataset
    (train_dataset, test_dataset),
        dataset_info = tfds.load('yelp_polarity_reviews',
            split=['train', 'test'], shuffle_files=True,
            with_info=True, as_supervised=True)
    ```

    We use the tf.load function to fetch datasets from TensorFlow datasets. Here, we specify our dataset, which is the Yelp Polarity reviews dataset. We also split our data into training and testing sets. We shuffle our data by setting shuffle to True, and we set with_info=True to ensure we can retrieve metadata of the dataset, which can be accessed using the dataset_info variable. We also set as_supervised=True; when we do this, it returns a tuple made up of the input and target rather than a dictionary. This way we can directly use the dataset with the fit method to train our models. We now have our training dataset as train_dataset and our testing set as test_dataset; both datasets are tf.data.Dataset objects. Let's proceed with some quick data exploration before we build our sentiment analysis models on our training data and evaluate them on the testing data.

3. Let's write some functions to enable us to explore our dataset:

```python
def get_reviews(dataset, num_samples=5):
    reviews = []
    for text, label in dataset.take(num_samples):
        reviews.append((text.numpy().decode('utf-8'),
            label.numpy()))
    return reviews

def dataset_insights(dataset, num_samples=2000):
    total_reviews = 0
    total_positive = 0
    total_negative = 0
    total_length = 0
    min_length = float('inf')
    max_length = 0

    for text, label in dataset.take(num_samples):
        total_reviews += 1
        review_length = len(text.numpy().decode(
            'utf-8').split())
        total_length += review_length
        if review_length < min_length:
            min_length = review_length
        if review_length > max_length:
            max_length = review_length
        if label.numpy() == 1:
            total_positive += 1
        else:
            total_negative += 1

    avg_length = total_length / total_reviews
    return min_length, max_length, avg_length,
        total_positive, total_negative

def plot_reviews(positive, negative):
    labels = ['Positive', 'Negative']
    counts = [positive, negative]
    plt.bar(labels, counts, color=['blue', 'red'])
    plt.xlabel('Review Type')
    plt.ylabel('Count')
    plt.title('Distribution of Reviews')
    plt.show()
```

We use the `get_reviews` function to examine reviews from either the training or testing sets. This function displays the specified number of reviews and their corresponding labels; by default, it displays the first five reviews. However, we can set this parameter to any number we want. The second function is the `dataset_insight` function – this function performs several analyses, such as extracting the shortest, longest, and average length of reviews. It also generates the total count of positive and negative reviews in the dataset. Because we are working with a large dataset, we set `dataset_insight` to explore the first 2,000 samples. If you increase the number of samples, it will take a long time to analyze the data. We pass the total positive and negative count of reviews into the `plot_reviews` function to give us a graphical distribution of the data.

4. Let's check out the first seven reviews in our training data:

```
# Check out some reviews
print("Training Set Reviews:")
train_reviews = get_reviews(train_dataset, 7)
for review, label in train_reviews:
    print(f"Label: {label}, Review: {review[:100]}")
```

When we run the code, it returns the top seven reviews. Also, we only return the first 100 characters of each review for brevity:

```
Training Set Reviews:
Label: 1, Review: The Groovy P. and I ventured to his old
stomping grounds for lunch today.  The '5 and Diner' on 16th...
Label: 0, Review: Mediocre burgers - if you are in the area and
want a fast food burger, Fatburger is  a better bet th...
Label: 0, Review: Not at all impressed...our server was not very
happy to be there...food was very sub-par and it was ...
Label: 0, Review: I wish I would have read Megan P's review
before I decided to cancel my dinner reservations because ...
Label: 1, Review: A large selection of food from all over the
world. Great atmosphere and ambiance.  Quality of food i...
Label: 1, Review: I know, I know a review for Subway, come
on.  But I have to say that the service at this subway is t...
Label: 1, Review: We came in for a pre-bachelor party madness
meal and I have to say it was one of the best dining exp...
```

5. Let's examine some important statistics about our training data:

```
min_length, max_length, avg_length, total_positive,
    total_negative = dataset_insights(train_dataset)

# Display the results
print(f"Shortest Review Length: {min_length}")
```

```
print(f"Longest Review Length: {max_length}")
print(f"Average Review Length: {avg_length:.2f}")
print(f"Total Positive Reviews: {total_positive}")
print(f"Total Negative Reviews: {total_negative}")
```

When we run the code, it returns the following:

```
Shortest Review Length: 1
Longest Review Length: 942
Average Review Length: 131.53
Total Positive Reviews: 1030
Total Negative Reviews: 970
```

6. Let's plot the distributions of our sampled training data:

```
plot_reviews(total_positive, total_negative)
```

Here, we call the `plot_reviews` function and pass in the total number of positive and negative reviews from our sampled training data. When we run the code, we get the plot shown in *Figure 10.4*.

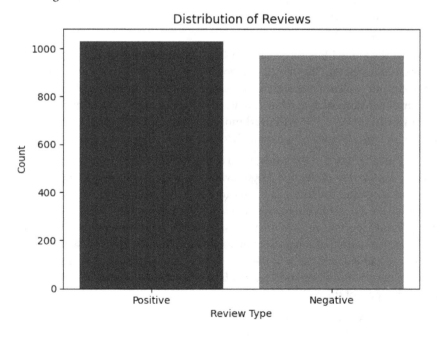

Figure 10.4 – A distribution of reviews from our sampled training data

From the sampled training dataset, we see that our reviews are finely balanced. Therefore, we can proceed to train our model on this dataset. Let's do that now.

7.   We define the key parameters for the tokenization, sequencing, and training processes:

```
# Define parameters
vocab_size = 10000
embedding_dim = 16
max_length = 132
trunc_type='post'
padding_type='post'
oov_tok = "<OOV>"
num_epochs = 10
# Build the Tokenizer
tokenizer = Tokenizer(num_words=vocab_size,
    oov_token=oov_tok)
```

We set our vocabulary size to 10,000. This means the tokenizer will focus on the top 10,000 words in our dataset. When selecting the vocabulary size, there is a need to strike a balance between computational efficiency and capturing word diversity present within the dataset under consideration. If we increase the vocab size, we are likely to capture more nuances that can enrich our model's understanding, but this will require more computational resources for training. Also, if we reduce the vocab size, training will be much faster; however, we will only capture a small portion of the linguistic variations present in our dataset.

Next, we set the embedding dimension to 16. This means each word will be represented by a 16-dimensional vector. The choice of embedding dimension is usually based on empirical testing. Here, our choice of embedding dimension was based on computational efficiency. If we use higher dimensions such as 64 or 128, we are likely to capture more nuanced relationships between words; however, we will need more computational resources for training. When working with large datasets, you may wish to use higher dimensions for better performance.

We set our max length to 132 words; we use this based on the average word length we got during our exploration of the first 2,000 reviews in our data. Reviews longer than 132 words will be truncated after the first 132 words are selected. Our choice of maximum length is to ensure we strike a decent compromise between computational efficiency and capturing the most important aspects of most of the reviews in our dataset. We set truncation and padding to post; this ensures that longer sentences are cut off at the end of a sequence and shorter sentences are padded with zeros at the end of the sequence. The key assumption here is that most of the important information we will find in a customer's review is likely to be found at the beginning part of the review.

Next, we set the OOV token to cater to OOV words that may occur in the test set but which the model did not see during training. Setting this parameter prevents our model from running into errors when handling unseen words. We also set the number of epochs that our model will train for to 10. Although we use 10 to test out our model, you may wish to train for longer and perhaps use callbacks to monitor the model's performance during training on a validation set.

With all our parameters defined, we can now instantiate our `Tokenizer` class, passing in `num_words` and `oov_token` as arguments.

8. To reduce the processing time, we will make use of the 20,000 samples for training:

```
# Fetch and decode the training data
train_text = []
train_label = []
for example in train_dataset.take(20000):
    text, label = example
    train_text.append(text.numpy().decode('utf-8'))
    train_label.append(label.numpy())

# Convert labels to numpy array
train_labels = np.array(train_label)

# Fit the tokenizer on the training texts
tokenizer.fit_on_texts(train_text)

# Get the word index from the tokenizer
word_index = tokenizer.word_index

# Convert texts to sequences
train_sequences = tokenizer.texts_to_sequences(
    train_text)
```

Here, we train our model with the first 20,000 samples from the Yelp Polarity training dataset. We collect these reviews and their corresponding labels, and since the data is in the form of bytes, we use UTF-8 encoding to decode the string, after which we append the text and their labels to their respective lists. We convert the list of labels for easy manipulation using NumPy. After this, we fit the tokenizer on our selected training data and convert the text to a sequence.

9. For testing purposes, we take 8,000 samples. The set of steps we carry out here is quite similar to those on the training set; however, we do not fit on text on the test set. This step is only for training purposes to help the neural network learn the word-to-index mapping in the training set:

```
# Fetch and decode the test data
test_text = []
test_label = []
for example in test_dataset.take(8000):
    text, label = example
    test_text.append(text.numpy().decode('utf-8'))
    test_label.append(label.numpy())

# Convert labels to numpy array
```

```
test_labels = np.array(test_label)
# Convert texts to sequences
test_sequences = tokenizer.texts_to_sequences(
    test_text)
```

We take the first 8,000 samples from our test dataset. It is important to use the same tokenizer that was used to fit our training data here. This ensures that the word index mapping learned by the tokenizer during training is applied to the test set, and words not learned in the training set are replaced with the OOV token.

10. The next step is to pad and truncate the sequences of integers representing the texts in the training and testing sets, ensuring that they all have the same length:

```
# Pad the sequences
train_padded = pad_sequences(train_sequences,
    maxlen=max_length, padding=padding_type,
    truncating=trunc_type)
test_padded = pad_sequences(test_sequences,
    maxlen=max_length, padding=padding_type,
    truncating=trunc_type)
```

The output returned, `train_padded` and `test_padded`, is NumPy arrays of shape (num_sequences and maxlen). Now, every sequence that makes up these arrays is of the same length.

11. We want to set up a validation set, which will help us track how our modeling process is going. To do this, we can use the `train_test_split` function from scikit-learn:

```
# Split the data into training and validation sets
train_padded, val_padded, train_labels,
    val_labels = train_test_split(train_padded,
        train_labels, test_size=0.2, random_state=42)
```

Here, we split the data into training and validation sets, with 20 percent set as the validation set.

12. Let's proceed to build our sentiment analysis model:

```
# Define the model
model = tf.keras.Sequential([
    tf.keras.layers.Embedding(vocab_size,
        embedding_dim, input_length=max_length),
    tf.keras.layers.GlobalAveragePooling1D(),
    tf.keras.layers.Dense(24, activation='relu'),
    tf.keras.layers.Dense(1, activation='sigmoid')
# because it's binary classification
])
```

We build our model with TensorFlow's Keras API. Note that we have a new layer called the embedding layer, which is used to represent words in a dense vector space. This layer takes in the vocabulary size, the embedding dimension, and the max length as its parameters. In this experiment, we are training our word embeddings as a part of our model. We can also train this layer independently for the purpose of learning word embeddings. This can come in handy when we intend to use the same word embeddings across multiple models. In *Chapter 11, NLP with TensorFlow*, we will see how to apply a pretrained embedding layer from TensorFlow Hub.

When we pass in a two-dimensional tensor of shape (`batch_size`, `input_length`), where each sample is an integer sequence, the embedding layer returns a three-dimensional tensor of shape (`batch_size`, `input_length`, `embedding_dim`). At the start of training, embedding vectors are randomly initialized. As we train the model, these vectors are adjusted, ensuring words with a similar context are clustered closely together within the embedding space. Instead of using discrete values, word embeddings make use of continuous values that our model can use to discern patterns and model intricate relationships with the data.

The `GlobalAveragePooling1D` layer is applied to reduce the dimensionality of our data; it applies an average pooling operation. For example, if we apply `GlobalAveragePooling1D` to a sequence of words, it will return a summarized, single vector as output that can be fed into our fully connected layers for classification. Because we are performing binary classification, we use one neuron in our output layer and a sigmoid as our activation function.

13. Now, we compile and fit our model. We pass in the loss as `binary_crossentropy` for the compilation step. We use Adam as our optimizer, and for our classification metric, we use accuracy:

```
# Compile the model
model.compile(loss='binary_crossentropy',
    optimizer='adam', metrics=['accuracy'])
```

14. We fit the model for 10 epochs using our training data (`train_padded`) and labels (`train_labels`) and use the validation data to track our experiment:

```
# Train the model
history = model.fit(train_padded, train_labels,
    epochs=num_epochs, validation_data=(val_padded,
        val_labels))
```

We report the results from the last 5 epochs:

```
Epoch 6/10
625/625 [==============================] - 4s 7ms/step - loss:
0.1293 - accuracy: 0.9551 - val_loss: 0.3149 - val_accuracy:
0.8875
Epoch 7/10
625/625 [==============================] - 4s 6ms/step - loss:
0.1116 - accuracy: 0.9638 - val_loss: 0.3330 - val_accuracy:
0.8880
```

```
Epoch 8/10
625/625 [==============================] - 5s 9ms/step - loss:
0.0960 - accuracy: 0.9697 - val_loss: 0.3703 - val_accuracy:
0.8813
Epoch 9/10
625/625 [==============================] - 4s 6ms/step - loss:
0.0828 - accuracy: 0.9751 - val_loss: 0.3885 - val_accuracy:
0.8796
Epoch 10/10
625/625 [==============================] - 4s 6ms/step - loss:
0.0727 - accuracy: 0.9786 - val_loss: 0.4258 - val_accuracy:
0.8783
```

The model reaches an accuracy of 0.9786 on training and a validation accuracy of 0.8783. This tells us there is an element of overfitting. Let's see how our model will do on unseen data. To do this, let's evaluate the model with our test data.

15. We use the `evaluate` function to evaluate the trained model:

```
# Evaluate the model on the test set
results = model.evaluate(test_padded, test_labels)

print("Test Loss: ", results[0])
print("Test Accuracy: ", results[1])
```

We pass in the test data (`test_padded`) and test labels (`test_labels`) and print out the loss and accuracy. The model reached an accuracy of 0.8783 on the test set.

16. It is good practice to plot the loss and accuracy curves during training and validation, as it provides us with valuable insights into the learning process of the model and its performance. To do this, let's construct a function called `plot_history`:

```
def plot_history(history):
    plt.figure(figsize=(12, 4))

    # Plot training & validation accuracy values
    plt.subplot(1, 2, 1)
    plt.plot(history.history['accuracy'])
    plt.plot(history.history['val_accuracy'])
    plt.title('Model accuracy')
    plt.ylabel('Accuracy')
    plt.xlabel('Epoch')
    plt.legend(['Train', 'Validation'],
        loc='upper left')

    # Plot training & validation loss values
    plt.subplot(1, 2, 2)
```

```
plt.plot(history.history['loss'])
plt.plot(history.history['val_loss'])
plt.title('Model loss')
plt.ylabel('Loss')
plt.xlabel('Epoch')
plt.legend(['Train', 'Validation'],
    loc='upper right')

plt.tight_layout()
plt.show()
```

This function takes in the `history` object and returns to us both the loss and accuracy curves. The `plot_history` function will create a figure with two subplots – the subplot on the left shows the training and validation accuracy per epoch, and the subplot on the right shows the training and validation loss per epoch.

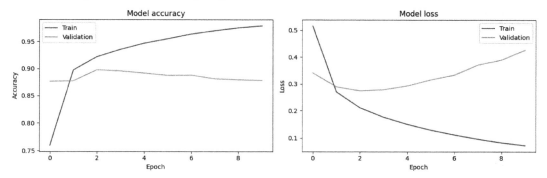

Figure 10.5 – The loss and accuracy curves

From the plots in *Figure 10.5*, we can see that the model's accuracy on training increases per epoch; however, the validation accuracy begins to fall slightly around the end of the first epoch. The training loss also falls steadily per epoch, while the validation loss rises steadily over each epoch, thus indicating overfitting.

17. Before we explore ways to fix overfitting in our case study, let's try four new sentences and see how our model fares on them:

```
# New sentences
new_sentences = ["The restaurant was absolutely fantastic. The
staff were kind and the food was delicious.",  # positive
    "I've had an incredible day at the beach, the weather was
beautiful.",  # positive
    "The movie was a big disappointment. I wouldn't recommend it
to anyone.",  # negative
    "I bought a new phone and it stopped working after a week.
Terrible product."]  # negative
```

18. In this example, we have the sentiments of each sentence indicated for reference. Let's see how our trained model will perform on each of these new sentences:

```
# Preprocess the sentences in the same way as the training data
new_sequences = tokenizer.texts_to_sequences(
    new_sentences)
new_padded = pad_sequences(new_sequences,
    maxlen=max_length, padding=padding_type,
    truncating=trunc_type)


# Use the model to predict the sentiment of the new sentences
predictions = model.predict(new_padded)


# Print out the sequences and the corresponding predictions
for i in range(len(new_sentences)):
    print("Sequence:", new_sequences[i])
    print("Predicted sentiment (
        probability):", predictions[i])
    if predictions[i] > 0.5:
        print("Interpretation: Positive sentiment")
    else:
        print("Interpretation: Negative sentiment")
    print("\n")
```

19. Let's print out the sequence corresponding to each sentence, along with the sentiment the model predicted:

```
1/1 [==============================] - 0s 21ms/step
Sequence: [2, 107, 7, 487, 533, 2, 123, 27, 290, 3, 2, 31, 7,
182]
Predicted sentiment (probability): [0.9689689]
Interpretation: Positive sentiment


Sequence: [112, 25, 60, 1251, 151, 26, 2, 3177, 2, 2079, 7, 634]
Predicted sentiment (probability): [0.9956489]
Interpretation: Positive sentiment


Sequence: [2, 1050, 7, 6, 221, 1174, 4, 454, 234, 9, 5, 528]
Predicted sentiment (probability): [0.43672907]
Interpretation: Negative sentiment


Sequence: [4, 764, 6, 161, 483, 3, 9, 695, 524, 83, 6, 393, 464,
1341]
Predicted sentiment (probability): [0.36306405]
Interpretation: Negative sentiment
```

Our sentiment analysis model was able to effectively predict the results correctly. What about if we want to visualize embeddings? TensorFlow has an embedding projector, which can be accessed at https://projector.tensorflow.org.

Embedding visualization

If we wish to visualize word embeddings from our trained model, we will need to extract the learned embeddings from the embedding layer and load them into the embedding projector provided by TensorBoard. Let's examine how we can do this:

1. Extract the embedding layer weights:

    ```
    weights = model.get_layer(
        'embedding').get_weights()[0]
    vocab = tokenizer.word_index
    print(weights.shape)
    # shape: (vocab_size, embedding_dim)
    ```

    The first step is to retrieve the learned weights from our embedding layer after training. Next, we obtain the word index mapping that was generated during the tokenization process. If we apply a print statement, we can see the vocabulary size and the embedding dimension.

2. Then, save the weights and vocabulary to disk. The TensorFlow Projector reads these file types and uses them to plot the vectors in 3D space, so we can visualize them:

    ```
    out_v = io.open('vectors.tsv', 'w', encoding='utf-8')
    out_m = io.open('metadata.tsv', 'w', encoding='utf-8')

    for word, index in vocab.items():
        if index < vocab_size:
            vec = weights[index]
            out_v.write('\t'.join([str(x) for x in vec]) + "\n")
            out_m.write(word + "\n")

    out_v.close()
    out_m.close()
    ```

3. The next step is to save the embedding vectors and the vocabulary (words) as two separate **tab-separated values (TSV)** files called vecs.tsv and meta.tsv, respectively. When we run this code block, we see that we have two new files in our Google Colab notebook, as shown in *Figure 10.6*.

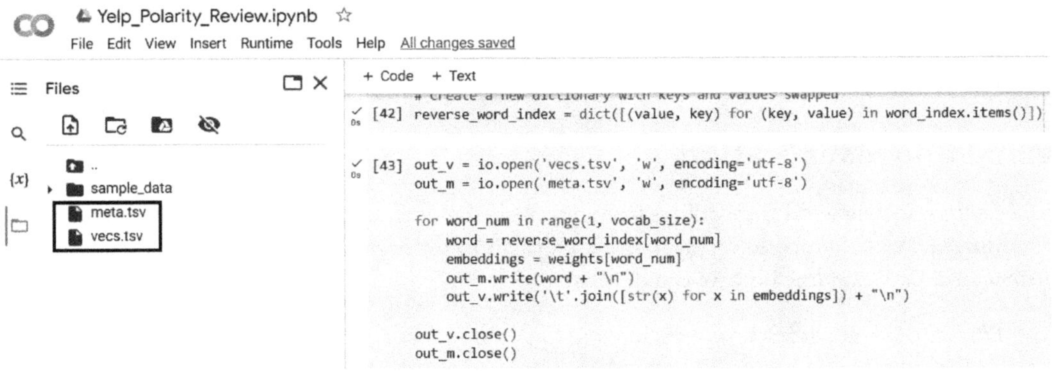

Figure 10.6 – A screenshot showing the meta and vecs files

4. Download the files locally:

```
try:
    from google.colab import files
    files.download('vectors.tsv')
    files.download('metadata.tsv')
except Exception:
    pass
```

To download the required files from Google Colab to our local machine, run this code block. Note that you need to move these files from your server to your local machine if you work in a cloud environment.

5. Visualize the embeddings. Open the embedding projector using this link: `https://projector.tensorflow.org/`. Then, you will have to click on the load button to load the `vectors.tsv` and `metadata.tsv` files you downloaded to your local machine. Once you successfully upload the files, the word embeddings will appear in 3D, as illustrated in *Figure 10.7*.

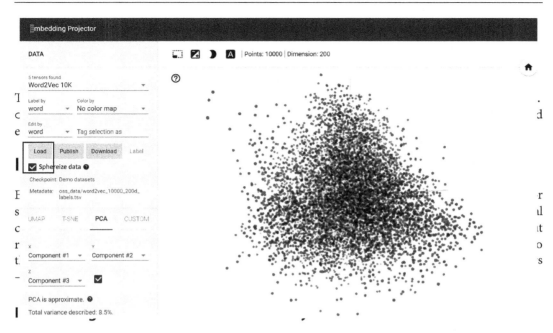

One hyperparameter we may consider changing is the size of the vocabulary. Increasing the vocabulary size empowers the model to learn more unique words from our dataset. Let's see how this will impact the performance of our base model. Here, we adjust `vocab_size` from `10000` to `20000`, while keeping the other hyperparameters constant:

```
# Increasing the vocab_size
vocab_size = 10000 #Change from 10000 to 20000
embedding_dim = 16
training_size = 20000
num_epochs=10

model_1, history_1 = sentiment_model(vocab_size,
    embedding_dim, training_size, num_epochs)
```

The model reaches a test accuracy of 0.8749 compared to 0.8783, which was achieved by our base model. Here, increasing `vocab_size` had no positive impact on the performance of our model.

When we use a larger vocabulary size, our model will learn more unique words, which could be a good idea, depending on the dataset and the use case. On the downside, more parameters and computational resource is required for us to efficiently train our model. Again, there is a greater risk of overfitting.

In light of these issues, it is important to strike the right balance by ensuring we have a large enough `vocab_size` to capture the nuances in our data, without introducing the risk of overfitting at the same time. One strategy is to set a minimum frequency threshold, such that rare words that may lead to overfitting are excluded from our vocabulary. Another idea we can try is to adjust the dimensions of the embedding. Let's discuss that next.

Adjusting the embedding dimension

The embedding dimension refers to the size of the vector space in which words are represented. A high-dimension embedding has the ability to capture more nuanced relationships between words. However, it also increases the model complexity and may lead to overfitting, especially when working with small datasets. Let's adjust embedding_dim from 16 to 32 while keeping other parameters constant and see what the impact will be on our experiment:

```
vocab_size = 10000
embedding_dim = 32 #Change from 16 to 32
train_size = 20000
num_epochs=10
model_2, history_2 = sentiment_model(vocab_size,
    embedding_dim, train_size, num_epochs)
```

In 10 epochs, our new model with a larger embedding dimension reached an accuracy of 0.8720 on the test set. This falls short of our baseline model. Here, we see firsthand that an increase in the embedding dimension doesn't always guarantee a better-performing model. When the embedding dimension is too small, it may fail to capture important relationships in our data. Conversely, an oversized embedding will lead to increased computation requirements and a greater risk of overfitting. It is important to note that a small embedding dimension suffices for simpler tasks or smaller datasets, while a larger embedding is an excellent choice for a large dataset. A pragmatic approach is to begin with a smaller embedding and gradually increase its size, while keeping an eye on the model's performance during each iteration. Usually, the performance will improve, and at some point, diminishing returns will set in. When this happens, we stop training. Now, we can collect more data, increase the number of samples, and see what our results look like.

Collecting more data

In *Chapter 8, Handling Overfitting*, we explored this option when handling overfitting. Collecting more data samples enables us to have a more diverse set of examples that our model can learn from. However, this process can be time-consuming. Also, more data may not help in cases where it is noisy or irrelevant. For our case study, let's increase the training data size from 20,000 samples to 40,000 samples:

```
vocab_size = 10000
embedding_dim = 16
train_size = 40000

model_3, history_3 = sentiment_model(vocab_size,
    embedding_dim, train_size)
```

After 10 epochs of training, we see that our model's performance improves to 0.8726 on our test set. Collecting more data can be a good strategy, as it may provide our model with a more diverse dataset; however, it didn't work in this instance. So, let's move on to other ideas; this time, let's try dropout regularization.

Dropout regularization

In *Chapter 8, Handling Overfitting*, we discussed dropout regularization, where we randomly dropped out a percentage of neurons from our model during training to break co-dependence among neurons in our model. Since we are dealing with a case of overfitting, let's try out this technique. To implement dropout in our model, we can add a dropout layer, as shown here:

```
model_4  = tf.keras.Sequential([
    tf.keras.layers.Embedding(vocab_size, embedding_dim,
    input_length=max_length),
    tf.keras.layers.GlobalAveragePooling1D(),
    tf.keras.layers.Dense(24, activation='relu'),
    tf.keras.layers.Dropout(0.5),
# Dropout layer with 50% dropout rate
    tf.keras.layers.Dense(1, activation='sigmoid')
])
# Compile the model
model_4.compile(loss='binary_crossentropy',
    optimizer='adam',metrics=['accuracy'])
```

Here, we set our dropout rate to 50 percent, meaning we turn off half of the neurons during training. Our new model achieves an accuracy of 0.8830, which is marginally better than our baseline model. Dropout can help to enhance the robustness of our model by preventing co-dependence between neurons in it. However, we must apply dropout with caution. If we drop out too many neurons, our model becomes too simple and begins to underfit because it is unable to capture the underlying patterns in our data. Also, if we apply a low dropout value to our model, we may not achieve the desired regularization effect we hope for. It is a good idea to experiment with different dropout values to find the best balance between model complexity and generalization. Now, let's try out different optimizers.

Trying a different optimizer

While Adam is a good general-purpose optimizer, you might find that a different optimizer, such as SGD or RMSprop, works better for your specific task. Different optimizers might work better, depending on the task at hand. Let's try out RMSprop for our use case and see how it fares:

```
# Initialize the optimizer
optimizer = tf.keras.optimizers.RMSprop(learning_rate=0.001)

# Compile the model
model_7.compile(loss='binary_crossentropy',
    optimizer=optimizer, metrics=['accuracy'])

# Train the model
history_7 = model_7.fit(train_padded, train_labels,
    epochs=num_epochs, validation_data=(val_padded,
        val_labels))
```

We achieved a test accuracy of 0.8920, beating our baseline model, using RMSprop as our optimizer. When choosing an appropriate optimizer, it is important to assess the key properties of your use case. For example, when working with large datasets, SGD is a more suitable choice than batch gradient descent, as SGD's use of mini-batches reduces the computational cost. This attribute is useful when working with limited computation resources. It's worth noting that if we have too many small batches while using SGD, it could lead to noisy updates; on the flip side, very large batch sizes could increase the computational cost.

Adam is an excellent default optimizer for many deep learning use cases, as it combines the advantages of RMSprop and Momentum; however, when we deal with simple convex problems such as linear regression, SGD proves to be a better choice, due to Adam's overcompensation in these scenarios. With this, we draw the curtain on this chapter.

Summary

In this chapter, we explored the foundations of NLP. We began by looking at how to handle real-world text data, and we explored some preprocessing ideas, using tools such as Beautiful Soup, requests, and regular expressions. Then, we unpacked various ideas, such as tokenization, sequencing, and the use of word embedding to transform text data into vector representations, which not only preserved the sequential order of text data but also captured the relationships between words. We took a step further by building a sentiment analysis classifier using the Yelp Polarity dataset from the TensorFlow dataset. Finally, we performed a series of experiments with different hyperparameters in a bid to improve our base model's performance and overcome overfitting.

In the next chapter, we will introduce **Recurrent Neural Networks** (**RNNs**) and see how they do things differently from the DNN we used in this chapter. We will put RNNs to the test as we will build a new classifier with them. We will also take things a step further by experimenting with pretrained embeddings, and finally, we will round off the NLP section by generating text in a fun exercise, using a dataset of children's stories. See you there.

Questions

Let's test what we have learned in this chapter.

1. Using the test notebook, load the IMDB dataset from TFDS.
2. Use a different embedding dimension and evaluate the model on the test set.
3. Use a different vocabulary size and evaluate the model on the test set.
4. Add more layers and evaluate the model on the test set.
5. Use your best model to make predictions on the sample sentences given.

Further reading

To learn more, you can check out the following resources:

- Kapoor, A., Gulli, A. and Pal, S. (2020) *Deep Learning with TensorFlow and Keras, Third Edition*. Packt Publishing Ltd.

- *Twitter Sentiment Classification using Distant Supervision* by Go et al. (2009)

- *Embedding Projector: Interactive Visualization and Interpretation of Embeddings* by Smilkov et al. (2016)

- *A Sensitivity Analysis of (and Practitioners' Guide to) Convolutional Neural Networks for Sentence Classification* by Zhang et al. (2016)

11
NLP with TensorFlow

Text data is inherently sequential, defined by the order in which words occur. Words follow one another, building upon previous ideas and shaping those to come. Understanding the sequence of words and the context in which they are applied is straightforward for humans. However, this poses a significant challenge to feed-forward networks such as **convolutional neural networks** (**CNNs**) and traditional **deep neural networks** (**DNNs**). These models treat text data as independent inputs; hence, they miss the interconnected nature and flow of language. For example, let's take the sentence "*The cat, which is a mammal, likes to chase mice*." Humans immediately recognize the relationship between the cat and mice, as we process the entire sentence as a whole and not individual units.

A **recurrent neural network** (**RNN**) is a type of neural network designed to handle sequential data such as text and time-series data. When working with text data, RNNs' memory enables them to recall earlier parts of a sequence, aiding them to understand the context in which words are used in a text. For example, with a sentence such as "*As an archaeologist, John loves discovering ancient artifacts,*" an RNN, in this context, can infer that an archaeologist will be excited about ancient artifacts, and in this sentence, the archaeologist is John.

In this chapter, we will begin to explore the world of RNNs and take a look under the hood to understand how the inner mechanisms of RNNs work together to maintain a form of memory. We will explore the pros and cons of RNNs when working with text data, after which we will switch our attention to investigating its variants, such as **long short-term memory** (**LSTM**) and **gated recurrent units** (**GRUs**). Next, we will use the knowledge we have gained to build a multiclass text classifier. We will then explore the power of transfer learning in the field of **natural language processing** (**NLP**). Here, we will see how to apply pretrained word embeddings to our workflow. To close this chapter, we will use RNNs to build a child story generator; here, we will see RNNs in action as they generate text data.

In this chapter, we will cover the following topics:

- The anatomy of RNNs
- Text classification with RNNs
- NLP with transfer learning
- Text generation

Understanding sequential data processing – from traditional neural networks to RNNs and LSTMs

In traditional neural networks, as we discussed earlier in this book, we see an arrangement of densely interconnected neurons, devoid of any form of memory. When we feed a sequence of data to these networks, it's an all-or-nothing transaction – the entire sequence is processed at once and converted into a singular vector representation. This approach is quite different from how humans process and comprehend text data. When we read, we naturally analyze text word by word, understanding that important words – those that have the power to shift the entire message of a sentence – can be positioned anywhere within it. For example, let's consider the sentence "*I loved the movie, despite some critics.*" Here, the word "*despite*" is pivotal, altering the direction of the sentiment expressed in the sentence.

RNNs don't just consider the value of individual words through embeddings; they also take into account the sequence or relative order of these words. This ordering of words gives them meaning and allows humans to effectively communicate with one another. RNNs are unique in their ability to retain context from one timestamp (or one word in the case of a sentence) to the next, thereby preserving the sequential coherence of the input. For example, in the sentence "*I visited Rome last year, and I found the Colosseum fascinating,*" an RNN would understand that "*the Colosseum*" relates to "*Rome*" because of the sequence of words. However, there's a catch – this context retention can fail in longer sentences where the distance between related words increases.

This is precisely where gated variants of RNNs such as LSTM networks come in. LSTMs are designed with a special "cell state" architecture that enables them to manage and retain information over longer sequences. So, even in a lengthy sentence such as "*I visited Rome last year, experienced the rich culture, enjoyed the delicious food, met wonderful people, and I found the Colosseum fascinating,*" an LSTM could still link "*the Colosseum*" with "*Rome,*" understanding the broader context despite the length and complexity of the sentence. We have only scratched the surface. Let's now examine the anatomy of these powerful networks. We will begin with RNNs.

The anatomy of RNNs

In the previous section, we talked about RNNs' ability to handle sequential data; let's drill down into how an RNN does this. The key differentiator between RNNs and feed-forward networks is their internal memory, as shown in *Figure 11.1*, which enables RNNs to process input sequences while retaining information from previous steps. This attribute empowers RNNs to suitably exploit the temporal dependencies in sequences such as text data.

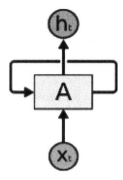

Figure 11.1 – The anatomy of an RNN

Figure 11.2 shows a clearer picture of an RNN and its inner workings. Here, we can see a series of interconnected units through which data flows in a sequential fashion, one element at a time. As each unit processes the input data, it sends the output to the next unit in a similar fashion to how feed-forward networks work. The key difference lies in the feedback loop, which equips RNNs with the memory of previous inputs, empowering them with the ability to comprehend entire sequences.

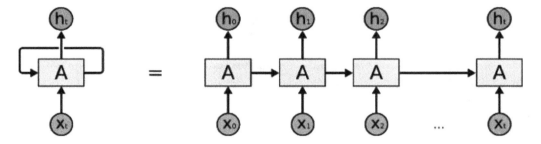

Figure 11.2 – An expanded view of an RNN showing its operation across multiple timesteps

Let's imagine we are dealing with sentences, and we want our RNN to learn something about their grammar. Each word in the sentence represents a time step, and at each one, the RNN considers the current word and also the "context" from the previous words (or steps). Let's go over a sample sentence. Let's say we have a sentence with five words – *"Barcelona is a nice city."* This sentence has five time steps, one for each word. At time step 1, we feed the word *"Barcelona"* into our RNN. The network learns something about this word (in reality, it would learn from the word's vector representation), produces an output, and also a hidden state capturing what it has learned, as illustrated in *Figure 11.2*. Now, we unroll the RNN to timestep 2. We input the next word, *"is,"* into our network, but we also input the hidden state from timestep 1. This hidden state represents the "memory" of the network, allowing the network to take into account what it has seen so far. The network produces a new output and a new hidden state.

This process continues, with the RNN unrolling further for each word in the sentence. At each time step, the network takes the current word and the hidden state from the previous time step as input, producing an output and a new hidden state. When you "unroll" an RNN in this way, it may look like a deep feed-forward network with shared weights across each layer (since each timestep uses the same underlying RNN cell for its operations), but it's more accurately a single network that's being applied to each timestep, passing along the hidden state as it goes. RNNs have the ability to learn from and remember sequences of arbitrary lengths; however, they do have their own limitations. One key issue with RNNs is their struggle with capturing long-term dependencies due to the vanishing gradient problem. This happens because the influence of time steps over future time steps can diminish over long sequences, as a result of repeated multiplications during backpropagation, which makes the gradients exceedingly small and thus harder to learn from. To address this, we will apply more advanced versions of RNNs such as LSTM and GRU networks. These architectures apply gating mechanisms to control the flow of information in the network and make it easier for the model to learn long-term dependencies. Let's examine these variants of RNNs.

Variants of RNNs – LSTM and GRU

Let's imagine we are working on a movie review classification project, and while inspecting our dataset, we find a sentence such as this one: "*The movie started off boring and slow, but it really picked up toward the end, and the climax was amazing.*" By examining the sentence, we see that the initial set of words used by the reviewer portrays a negative sentiment by using words such as "slow" and "boring," but the sentiment takes a shift to a more positive one with the use of phrases such as "picked up" and "climax was amazing." If we use a simple RNN for this task, due to its inherent limitation to retain information over longer sequences, it may misclassify the sentence by attaching undue importance to the earlier negative tone of the review.

Conversely, LSTMs and GRUs are designed to handle long-term dependencies, which makes them effective in not just capturing the change in sentiment but also for other NLP tasks, such as machine translation, text summarization, and question answering, where they outshine their simpler counterparts.

LSTMs

LSTMs are a specialized type of RNN designed to address the vanishing gradient problem, enabling LSTMs to effectively handle long-term dependencies in sequential data. To resolve this issue LSTM introduced a new structure called a memory cell, which essentially acts as an information carrier with the ability to preserve information over an extended period. Unlike standard RNNs, which feed information from one step to the next and tend to lose it over time, an LSTM with the aid of its memory cell can store and retrieve information from any point in the input sequence, irrespective of its length. Let's examine how an LSTM decides what information it will store in its memory cell. An LSTM is made up of four main components, as shown in *Figure 10.3*:

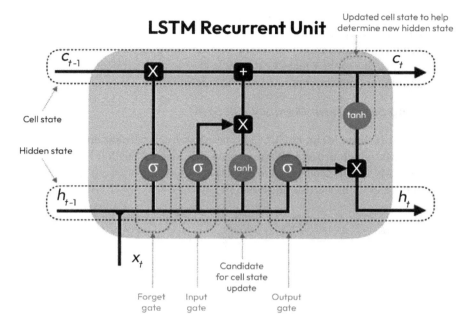

10.3 – LSTM architecture

These components allow LSTMs to store and access information over long sequences. Let's look at each of the components:

- **Input gate**: The input gate decides what new information will be stored in the memory cell. It is made up of a sigmoid and a tanh layer. The sigmoid layer produces output values between zero and one, representing the importance level of each value in the input, where zero means "not important at all" and one represents "very important." The tanh layer generates a set of candidate values that can be added to the state, essentially suggesting what new information should be stored in the memory cell. The outputs of both layers are merged by performing element-wise multiplication. This element-wise operation produces an input modulation gate that effectively filters the new candidate values, by deciding which information is important enough to be stored in the memory cell.

- **Forget gate**: This gate decides what information will be retained and what information will be discarded. It uses a sigmoid layer to return output values between zero and one. If a unit in the forget gate returns an output value close to zero, the LSTM will remove the information in the corresponding unit of the cell state.

- **Output gate**: The output gate determines what the next hidden state should be. Like the other gates, it also uses a sigmoid function to decide which parts of the cell state make it to the output.

- **Cell state**: The cell state is the "memory" of the LSTM cell. It is updated based on the output from the forget and input gates. It can remember information for use later in the sequence.

The gate mechanisms are paramount because they allow the LSTM to automatically learn appropriate context-dependent ways to read, write, and reset cells in the memory. These capabilities enable LSTMs to handle longer sequences, making them particularly useful for many complex sequential tasks where standard RNNs fall short, due to their inability to handle long-term dependencies, such as machine translation, text generation, time-series prediction, and video analysis.

Bidirectional Long Short-Term Memory (BiLSTM)

Bidirectional Long Short-Term Memory (**BiLSTM**) is an extension of traditional LSTM networks. However, unlike LSTMs, which process information in a sequential fashion from start to finish, BiLSTMs run two LSTMs simultaneously – one processes sequential data from the start to the end and the other from the end to the start, as illustrated in *Figure 11.4*.

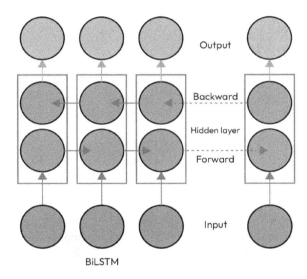

Figure 11.4 – The information flow in a BiLSTM

By doing this, BiLSTMs can capture both the past and future context of each data point in the sequence. Because of BiLSTMs' ability to comprehend the context from both directions in a sequence of data, they are well suited for tasks such as text generation, text classification, sentiment analysis, and machine translation. Now, let's examine GRUs.

GRUs

In 2014, *Cho et al.* introduced the **GRU** architecture as a viable alternative to LSTMs. GRUs were designed to achieve two primary goals – one was to overcome the vanishing gradient issues that plagued traditional RNNs, and the other was to streamline LSTM architecture for increased computational efficiency while maintaining the ability to model long-term dependencies. Structurally, the GRU has two primary gates, as illustrated in *Figure 11.5*.

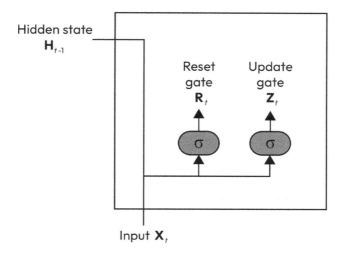

Figure 11.5 – A GRU's architecture

One key difference between the GRU and LSTM is the absence of a separate cell state in GRUs; instead, they use a hidden state to transfer and manipulate information as well as streamline its computational needs.

GRUs have two main gates, the update gate and the reset gate. Let's examine them:

- **Update gate**: The update gate condenses the input and forget gates in LSTMs. It determines how much of the past information needs to be carried forward to the current state and which information needs to be discarded.

- **Reset gate**: This gate defines how much of the past information should be forgotten. It helps the model evaluate the relative importance of new input against past memory.

Along with these two primary gates, a GRU introduces a "candidate hidden state." This candidate hidden state combines the new input and the previous hidden state, and by doing so, it develops a preliminary version of the hidden state for the current time step. This candidate then plays an important role in determining the final hidden state, ensuring that the GRU retains relevant context from the past while accommodating new information. When deciding between LSTMs and GRUs, the choice is largely dependent on the specific application and the computational resources available. For some applications, the increased computational efficiency of GRU is more appealing. For example, in real-time processing, such as text-to-speech, or when working on tasks with short sequences, such as sentiment analysis of tweets, GRUs could prove to be an excellent choice in comparison to LSTMs.

We have provided a high-level discussion on RNNs and their variants. Let's now proceed to apply these new architectures to a real-world use case. Will they outperform the standard DNNs or CNNs? Let's find out in a text classification case study.

Text classification using the AG News dataset – a comparative study

The AG News dataset is a collection of more than 1 million news articles, collected from over 2,000 news sources by a news search engine called ComeToMyHead. The dataset is distributed across four categories – namely, world, sports, business, and science and technology – and it is available on **TensorFlow Datasets** (**TFDS**). The dataset is made up of 120,000 training samples (30,000 from each category), and the test set contains 7,600 examples.

> **Note**
>
> This experiment may take about an hour to run, due to the size of the dataset and the number of models; hence, it is important to ensure your notebook is GPU-enabled. Again, you could take a smaller subset to ensure your experiments run much faster.

Let's start building our model:

1. We will begin by loading the necessary libraries for this experiment:

    ```
    import pandas as pd
    import tensorflow_datasets as tfds
    import tensorflow as tf
    from tensorflow.keras.preprocessing.text import Tokenizer
    from tensorflow.keras.preprocessing.sequence import pad_
    sequences
    from tensorflow.keras.utils import to_categorical
    import tensorflow_hub as hub
    from sklearn.model_selection import train_test_split
    from tensorflow.keras.models import Sequential
    from tensorflow.keras.layers import Embedding, SimpleRNN,
    Conv1D, GlobalMaxPooling1D, LSTM, GRU, Bidirectional, Dense,
    Flatten
    ```

    These imports form the building blocks, enabling us to solve this text classification problem.

2. Then, we load the AG News dataset from TFDS:

    ```
    # Load the dataset
    dataset, info = tfds.load('ag_news_subset',
        with_info=True, as_supervised=True)
    train_dataset, test_dataset = dataset['train'],
        dataset['test']
    ```

We use this code to load our dataset from TFDS – the `tfds.load` function fetches and loads the AG News dataset. We set the `with_info` argument to `True`; this ensures the metadata of our dataset, such as the total number of samples and the version, is also collected. This metadata information is stored in the `info` variable. We also set `as_supervised` to `True`; we do this to ensure that data is loaded in input and label pairs, where the input is the news article and the label is the corresponding category. Then, we split the data into a training set and a test set.

3. Now, we need to prepare our data for modeling:

```
# Tokenize and pad the sequences
tokenizer = Tokenizer(num_words=20000,
    oov_token="<OOV>")
train_texts = [x[0].numpy().decode(
    'utf-8') for x in train_dataset]
tokenizer.fit_on_texts(train_texts)
sequences = tokenizer.texts_to_sequences(train_texts)
sequences = pad_sequences(sequences, padding='post')
```

Here, we perform data preparatory steps such as tokenization, sequencing, and padding, using TensorFlow's Keras API. We initialize the tokenizer, which we use to convert our data from text to a sequence of integers. We set the `num_words` parameter to `20000`. This means we will only consider the top 20,000 occurring words in our dataset for tokenization; less frequently occurring words below this limit will be ignored. We set the `oov_token="<OOV>"` parameter to ensure we cater to unseen words that we may encounter during model inferencing.

Then, we extract the training data and store it in the `train_texts` variable. We tokenize and transform our data into a sequence of integers by mapping numerical values to tokens, using the `fit_on_texts` and `texts_to_sequences()` methods respectively. We apply padding to each sequence to ensure that the data we will input into our models is of a consistent shape. We set `padding` to `post`; this will ensure padding is applied at the end of a sequence. We now have our data in a well-structured format, which we will feed into our deep-learning models for text classification shortly.

4. Before we start modeling, we want to split our data into training and validation sets. We do this by splitting our training set into 80 percent for training and 20 percent for validation:

```
# Convert labels to one-hot encoding
train_labels = [label.numpy() for _, label in train_dataset]
train_labels = to_categorical(train_labels,
    num_classes=4)4  # assuming 4 classes

# Split the training set into training and validation sets
train_sequences, val_sequences, train_labels,
    val_labels = train_test_split(sequences,
        train_labels, test_size=0.2)
```

We convert our labels to one-hot encoding vectors, after which we split our training data using the `train_test_split` function from scikit-learn. We set our `test_size` to `0.2`; this means we will have 80 percent of our data for training and the remaining 20 percent for validation purposes.

5.   Let's set the `vocab_size`, `embedding_dim`, and `max_length` parameters:

```
vocab_size=20000
embedding_dim =64
max_length=sequences.shape[1]
```

We set `vocab_size` and `embedding_dim` to `20000` and `64`, respectively. When selecting your `vocab_size`, it is important to strike a good balance between computation efficiency, model complexity, and the ability to capture language nuances, while we use our embedding dimension to represent each word in our vocabulary by a 64-dimensional vector. The `max_length` parameter is set to match the longest tokenized and padded sequence in our data.

6.   We begin building our models, starting with a DNN:

```
# Define the DNN model
model_dnn = Sequential([
    Embedding(vocab_size, embedding_dim,
        input_length=max_length),
    Flatten(),
    tf.keras.layers.Dense(64, activation='relu'),
    Dense(16, activation='relu'),
    Dense(4, activation='softmax')
])
```

Using the `Sequential` API from TensorFlow, we build a DNN made up of an embedding layer, a flatten layer, two hidden layers, and an output layer for multiclass classification.

7.   Then, we build a CNN architecture:

```
# Define the CNN model
model_cnn = Sequential([
    Embedding(vocab_size, embedding_dim,
        input_length=max_length),
    Conv1D(128, 5, activation='relu'),
    GlobalMaxPooling1D(),
    tf.keras.layers.Dense(64, activation='relu'),
    Dense(4, activation='softmax')
])
```

We use a `Conv1D` layer made up of 128 filters (feature detectors) and a kernel size of 5; this means it will consider five words at a time. Our architecture uses `GlobalMaxPooling1D` to downsample the output of the convolutional layer to the most significant features. We feed the output of the pooling layer into a fully connected layer for classification.

8. Then, we build an LSTM model:

```
# Define the LSTM model
model_lstm = Sequential([
    Embedding(vocab_size, embedding_dim,
        input_length=max_length),
    LSTM(32, return_sequences=True),
    LSTM(32),
    tf.keras.layers.Dense(64, activation='relu'),
    Dense(4, activation='softmax')
])
```

Our LSTM architecture is made up of two LSTM layers made up of 32 units each. In the first LSTM layer, we set `return_sequences` to `True`; this allows the first LSTM layer to pass the complete sequence it received as output to the next LSTM layer. The idea here is to allow the second LSTM layer access to the context of the entire sequence; this equips it with the ability to better understand and capture dependencies across the entire sequence. We then feed the output of the second LSTM layer into the fully connected layers to classify our data.

9. For our final model, let's use a bidirectional LSTM:

```
# Define the BiLSTM model
model_BiLSTM = Sequential([
    Embedding(vocab_size, embedding_dim,
        input_length=max_length),
    Bidirectional(LSTM(32, return_sequences=True)),
    Bidirectional(LSTM(16)),
    tf.keras.layers.Dense(64, activation='relu'),
    Dense(4, activation='softmax')])
```

Here, instead of the LSTM layers, two bidirectional LSTM layers are added. Note that the first layer also has `return_sequences=True` to return the full outputs to the next layer. Using a bidirectional wrapper allows each LSTM layer access to both past and future context when processing each element of the input sequence, providing additional contextual information when compared with a unidirectional LSTM.

Stacking BiLSTM layers can help us build higher-level representations of the full sequence. The first BiLSTM extracts features by looking at the text from both directions while preserving the entire sequence. The second BiLSTM can then build on those features by further processing them. The final classification is carried out by the output layer in the fully connected layer. Our experimental models are now all set up, so let's proceed with compiling and fitting them next.

10. Let's compile and fit all the models we have built so far:

```
models = [model_cnn, model_dnn, model_lstm,
    model_BiLSTM]

for model in models:
    model.compile(loss='categorical_crossentropy',
        optimizer='adam', metrics=['accuracy'])
    model.fit(train_sequences, train_labels,epochs=10,
        validation_data=(val_sequences, val_labels),
        verbose=False
```

We use a `for` loop to compile and fit all four models. We set `verbose` to `False`; this way, we don't print the training information. We train for 10 epochs. Do expect this step to take a while, as we have a massive dataset and are experimenting with four models.

11. Let's evaluate our model on unseen data:

```
# Evaluate the model
test_texts = [x[0].numpy().decode(
    'utf-8') for x in test_dataset]
test_sequences = tokenizer.texts_to_sequences(
    test_texts)
test_sequences = pad_sequences(test_sequences,
    padding='post', maxlen=sequences.shape[1])

test_labels = [label.numpy() for _, label in test_dataset]
test_labels = to_categorical(test_labels,
    num_classes=4)

model_names = ["Model_CNN", "Model_DNN", "Model_LSTM",
    "Model_BiLSTM"]

for i, model in enumerate(models):
    loss, accuracy = model.evaluate(test_sequences,
        test_labels)

    print("Model Evaluation -", model_names[i])
    print("Loss:", loss)
    print("Accuracy:", accuracy)
    print()
```

To evaluate our model, we need to prepare our test data in the right fashion. We first extract our text data from our `test_dataset`, after which we tokenize the text using the tokenizer from our training process. The tokenized text is then converted into a sequence of integers, and padding is applied to ensure all sequences are of the same length as the longest sequence in our training data. Just like we did during training, we also one-hot-encode our test labels, and then we apply a `for` loop to iterate over each individual model, generating the test loss and accuracy for all our models. The output is as follows:

```
238/238 [==============================] - 1s 4ms/step - loss:
0.7756 - accuracy: 0.8989
Model Evaluation - Model_CNN
Loss: 0.7755934000015259
Accuracy: 0.8989473581314087

238/238 [==============================] - 1s 2ms/step - loss:
0.7091 - accuracy: 0.8896
Model Evaluation - Model_DNN
Loss: 0.7091193199157715
Accuracy: 0.8896052837371826

238/238 [==============================] - 2s 7ms/step - loss:
0.3211 - accuracy: 0.9008
Model Evaluation - Model_LSTM
Loss: 0.32113003730773926
Accuracy: 0.9007894992828369

238/238 [==============================] - 4s 10ms/step - loss:
0.5618 - accuracy: 0.8916
Model Evaluation - Model_BiLSTM
Loss: 0.5618014335632324
Accuracy: 0.8915789723396301
```

From our returned results, we can see that our LSTM model achieved the highest accuracy (90.08%); other models performed quite well too. We can take this performance as a good starting point; we can also apply some of the ideas we used in *Chapter 8, Handling Overfitting,* and *Chapter 10, Introduction to Natural Language Processing,* to improve our results here.

In *Chapter 10, Introduction to Natural Language Processing,* we talked about pretrained embeddings. These embeddings are trained on a large corpus of text data. Let's see how we can leverage them; perhaps they can help us achieve a better result in this case.

Using pretrained embeddings

In *Chapter 9, Transfer Learning*, we explored the concept of transfer learning. Here, we will revisit this concept as it relates to word embeddings. In all the models we have built up so far, we trained our word embeddings from scratch. Now, we will examine how to leverage pretrained embeddings that have been trained on massive amounts of text data, such as Word2Vec, GloVe, and FastText. Using these embeddings can be advantageous for two reasons:

- Firstly, they are already trained on a massive and diverse set of data, so they have a rich understanding of language.

- Secondly, the training process is much faster, since we will skip training our own word embeddings from scratch. Instead, we can build our models on the information packed in these embeddings, focusing on the task at hand.

It is important to note that using pretrained embeddings isn't always the right choice. For example, if you work on niche-based text data such as medical or legal data, sectors that apply a lot of domain-specific terminology may be underrepresented. When we use a pretrained embedding blindly for these use cases, they may lead to suboptimal performance. In these types of scenarios, you can either train your own embedding, which comes with an increased computational cost, or use a more balanced approach and fine-tune pretrained embeddings on your data. Let's see how we can apply pretrained embeddings in our workflow.

Text classification using pretrained embedding

To follow this experiment, you will need to use the second notebook in this chapter's GitHub repository called `modelling with pretrained embeddings`. We will continue with the same dataset. This time, we will focus on using our best model with the GloVe pretrained embedding. We will use our best model (LSTM) from our initial round of experiments. Let's start:

1. We will begin by importing the necessary libraries for this experiment:

    ```
    import numpy as np
    import tensorflow as tf
    from tensorflow.keras.models import Sequential
    from tensorflow.keras.layers import Embedding, LSTM, Dense,
    Flatten
    from tensorflow.keras.preprocessing.text import Tokenizer
    from tensorflow.keras.preprocessing.sequence import pad_
    sequences
    import tensorflow_datasets as tfds
    ```

    Once we import our libraries, we will download our pretrained embeddings.

2. Run the following commands to download the pretrained embeddings:

```
!wget http://nlp.stanford.edu/data/glove.6B.zip
!unzip glove.6B.zip -d glove.6B
```

We will download the GloVe 6B embedding files from the Stanford NLP website into our Colab notebooks, using the `wget` command. Then, we unzip the compressed files. We can see that this file contains different pretrained embeddings, as shown in *Figure 11.6*:

Figure 11.6 – The directory displaying the GloVe 6B embeddings files

The GloVe 6B embedding is made up of a 6 billion-token word embedding, trained by researchers at Stanford University and made publicly available to us all. For computational reasons, we will use the 50-dimensional vectors. You may wish to try out higher dimensions for richer representations, especially when working on tasks that require you to capture more complex semantic relationships, but also be mindful of the compute required.

3. Then, we load the AG News dataset:

```
dataset, info = tfds.load('ag_news_subset',
    with_info=True, as_supervised=True)
train_dataset, test_dataset = dataset['train'],
    dataset['test']
```

We load our dataset and split it into training and testing sets.

4. Then, we tokenize and sequence our training set:

```
tokenizer = Tokenizer(num_words=20000,
    oov_token="<OOV>")
train_texts = [x[0].numpy().decode(
    'utf-8') for x in train_dataset]
tokenizer.fit_on_texts(train_texts)
train_sequences = tokenizer.texts_to_sequences(
```

```
        train_texts)
train_sequences = pad_sequences(train_sequences,
    padding='post')
max_length = train_sequences.shape[1]
```

We prepare our data for modeling, just like our last experiment. We set the vocabulary size to 2,000 words and use OOV for out-of-vocabulary words. Then, we tokenize and pad the data to ensure consistency in the length of it.

5.   Then, we process our test data:

```
test_texts = [x[0].numpy().decode(
    'utf-8') for x in test_dataset]
test_sequences = tokenizer.texts_to_sequences(
    test_texts)
test_sequences = pad_sequences(test_sequences,
    padding='post', maxlen=max_length)
```

We process our testing data in a similar fashion to our training data. However, it is important to note that we do not apply `fit_on_texts` to our test set, ensuring that the tokenizer remains the same as the training set.

6.   Then, we set the embedding parameters:

```
vocab_size = len(tokenizer.word_index) + 1
embedding_dim = 50
```

We define the size of our vocabulary and we set the dimensionality of our embeddings to 50. We use this because we are working with 50d pretrained embeddings.

7.   Apply the pretrained word embeddings:

```
# Download GloVe embeddings and prepare embedding matrix
with open('/content/glove.6B/glove.6B.50d.txt', 'r',
encoding='utf-8') as f:
    for line in f:
        values = line.split()
        word = values[0]
        if word in tokenizer.word_index:
            idx = tokenizer.word_index[word]
            embedding_matrix[idx] = np.array(
                values[1:], dtype=np.float32)
```

We access the `glove.6B.50d.txt` file and read it line by line. Each line contains a word and its corresponding embeddings. We cross-match words in the GloVe file with those in our own vocabulary, constructed with the Keras tokenizer. If there is a match, we take the corresponding word index from our own vocabulary and update our initially zero-initialized embedding matrix at that index with the GloVe embeddings. Conversely, words that do not match will remain as zero vectors in the matrix. We will use this embedding matrix to initialize the weight of our embedding layer shortly. We use the file path to the 50d embeddings, as shown in *Figure 11.6*. You can do this by right-clicking on the specified file and copying the file path.

8. Let's build, compile, and train our LSTM model:

```
model_lstm = Sequential([
    Embedding(vocab_size, embedding_dim,
        input_length=max_length,
        weights=[embedding_matrix], trainable=False),
    LSTM(32, return_sequences=True),
    LSTM(32),
    Dense(64, activation='relu'),
    Dense(4, activation='softmax')
])

model_lstm.compile(optimizer='adam',
    loss='categorical_crossentropy',
    metrics=['accuracy'])

# Convert labels to one-hot encoding
train_labels = tf.keras.utils.to_categorical(
    [label.numpy() for _, label in train_dataset])
test_labels = tf.keras.utils.to_categorical(
    [label.numpy() for _, label in test_dataset])

model_lstm.fit(train_sequences, train_labels,
    epochs=10, validation_split=0.2)
```

While building our model, the key difference from our previous experiment is the embedding layer initialization. Here, we leverage our pretrained embedding matrix, and to ensure the weights remain unchanged, we set the trainable parameter to `false`. Everything else in our model's architecture remains the same. Then, we compile and fit our model for 10 epochs.

9. Finally, we evaluate our model:

```
loss, accuracy = model_lstm.evaluate(test_sequences,
    test_labels)
print("Loss:", loss)
print("Accuracy:", accuracy)
```

We reached an accuracy of 89% on our test set, although we did not outperform our best model when we didn't apply pretrained embeddings. Perhaps you may want to try out larger embedding dimensions from `glove6B` or other word embeddings to improve our result here. That would be a good exercise and is well encouraged.

Now, it is time to move on to another exciting topic – text generation with LSTM.

Using LSTMs to generate text

We have explored LSTMs in text classification. Now, we will look at how to generate text that you would see in a novel, blog post, or children's storybook, ensuring that is coherent and consistent with what we expect from these types of texts. LSTMs prove useful here, due to their ability to capture and remember intricate patterns for long sequences. When we train an LSTM on a large volume of text data, we allow it to learn the linguistic structure, style, and nuances. It can apply this to generate new sentences in line with the style and approach of the training set.

Let's imagine we are playing a word prediction game with our friends. The goal is to coin a story in which each friend comes up with a word to continue the story. To begin, we have a set of words, which we will call the seed, to set the tone for our story. From the seed sentence, each friend contributes a subsequent word until we have a complete story. We can also apply this idea to LSTMs – we feed a seed sentence into our model, and then we ask it to predict the next word, just like our word prediction game. This time, however, the game is played only by the LSTM, and we just need to specify the number of words it will play for. The result of each round of the game will serve as input to the LSTM for the next round, until we achieve our specified number of words.

The question that comes to mind is, how does an LSTM know what word to predict next? This is where the concept of **windowing** comes into play. Let's say we have a sample sentence such as "*I once had a dog called Jack.*" When we apply windowing with a window size of four to this sentence, we will have the following:

- "*I once had a*"
- "*once had a dog*"
- "*had a dog called*"
- "*a dog called Jack*"

We can now split each of these sentences into input-output pairs. For example, in the first sentence, we will have "*I once had*" as the input, and the output will be "*a.*" We will apply the same approach to all the other sentences, thus giving us the following input-output pairs:

- (["*I*", "*once*", "*had*"], "*a*")
- (["*once*", "*had*", "*a*"], "*dog*")
- (["*had*", "*a*", "*dog*"], "*called*")
- (["*a*", "*dog*", "*called*"], "*Jack*")

By using windowing, our LSTM focuses on the most recent set of words, which often holds the most relevant information to predict the next word in our text. Also, when we work with smaller, fixed-length sequences, it streamlines our training process and also optimizes our memory usage. Let's see how we can apply this idea in our next case study.

Story generation using LSTMs

In this case study, imagine you are a new NLP engineer working for a London-based start-up called Readrly. Your job is to build an AI storyteller for the company. You have been provided with a training dataset called `stories.txt`, which contains 30 sample stories. Your job is to train an LSTM to generate exciting children's stories. Let's return to our notebook and see how to make this happen:

1. As we did previously, we will begin by importing all the libraries required for this task:

    ```
    import tensorflow as tf
    from tensorflow.keras.preprocessing.text import Tokenizer
    from tensorflow.keras.preprocessing.sequence import pad_
    sequences
    from tensorflow.keras.models import Sequential
    from tensorflow.keras.layers import Embedding, LSTM, Dense,
    Bidirectional
    import numpy as np
    ```

2. Then, we load the `stories.txt` dataset:

    ```
    text = open('stories.txt').read().lower()
    ```

    We read our story file and convert all the contents to lowercase to ensure our data is in a consistent form, avoiding duplicate tokens generated from capitalized and non-capitalized versions of the same word.

    > **Note**
    >
    > This step helps us to reduce the vocabulary size by removing semantic nuances created by capitalization. However, this step should be carried out judiciously, as it may have a negative effect – for example, when we use the word "*march*." If we lowercase the word, it will denote some form of walking, while with a capital M, it refers to the month of the year.

3. Tokenize the text:

    ```
    tokenizer = Tokenizer()
    tokenizer.fit_on_texts([text])
    total_words = len(tokenizer.word_index) + 1
    ```

    We calculate the total number of unique words in our vocabulary. We add one to account for out-of-vocabulary words.

4. Then, we convert the text to sequences:

    ```
    input_sequences = []
    for line in text.split('\n'):
        token_list = tokenizer.texts_to_sequences(
            [line])[0]
    ```

```
for i in range(1, len(token_list)):
    n_gram_sequence = token_list[:i+1]
    input_sequences.append(n_gram_sequence)
```

In this step, we develop a dataset made up of *n*-gram sequences, where each entry in the "input sequence" is a sequence of words (that is, word numbers) that appear in the text. For every sequence of *n* words, *n*-1 words will be our input features, and the *n*-th word is the label that our model tries to predict.

Note

An n-gram is a sequence of n words from a given text.

For example, let's say we have a sample sentence such as "*The dog played with the cat.*" Using a 2-gram (bigram) model on a sentence would split it into the following bigrams:

("*The*", "*dog*"), ("*dog*", "*played*"), ("*played*", "*with*"), ("*with*", "*the*"), ("*the*", "*cat*")

Alternatively, if we use a 3-gram (trigram) model, it would split the sentence into trigrams:

("*The*", "*dog*", "*played*"), ("*dog*", "*played*", "*with*"), ("*played*", "*with*", "*the*"), ("*with*", "*the*", "*cat*")

N-gram models are common in NLP for text prediction, spelling correction, language modeling, and feature extraction.

---

5. Then, we pad the sequences:

```
max_sequence_len = max([len(x) for x in input_sequences])
input_sequences = np.array(pad_sequences(
    input_sequences, maxlen=max_sequence_len,
    padding='pre'))
```

We use padding to ensure our input data is of a consistent format. We set `padding` to `pre` to ensure that our LSTM captures the most recent words at the end of each sequence; these words are relevant to predict the next word in the sequence.

6. We now split the sequences into features and labels:

```
predictors, label = input_sequences[:,:-1],
    input_sequences[:,-1]
label = tf.keras.utils.to_categorical(label,
    num_classes=total_words)
```

Here, we one-hot-encode our labels to represent them as vectors. Then, we build our model.

7. Create the model:

```
model = Sequential([
    Embedding(total_words, 200,
        input_length=max_sequence_len-1),
```

```
    Bidirectional(LSTM(200)),
    Dense(total_words, activation='softmax')
])
model.compile(loss='categorical_crossentropy',
    optimizer='adam', metrics=['accuracy'])
history = model.fit(predictors, label, epochs=300,
    verbose=0)
```

We build a text generation model using a bidirectional LSTM, as it has the ability to capture both past and future data points. Our embedding layer transforms each word's numerical representation into a dense vector, with a dimension of 200 each. The next layer is the bidirectional LSTM layer with 200 units, after which we feed the output data into the dense layers. Then, we compile and fit the model for 300 epochs.

8. Create a function to make the predictions:

```
def generate_text(seed_text, next_words, model, max_sequence_
len):
    for _ in range(next_words):
        token_list = tokenizer.texts_to_sequences(
            [seed_text])[0]
        token_list = pad_sequences([token_list],
            maxlen=max_sequence_len-1, padding='pre')

        # Get the predictions
        predictions = model.predict(token_list)

        # Get the index with the maximum prediction value
        predicted = np.argmax(predictions)

        output_word = ""
        for word,index in tokenizer.word_index.items():
            if index == predicted:
                output_word = word
                break

        seed_text += " " + output_word

    return seed_text
```

We construct the `story_generator` function. We start with the seed text, which we use to prompt the model to generate more text for our children's stories. The seed text is transformed into its tokenized form, after which it is padded at the beginning to match the expected length of the input sequence, `max_sequence_len-1`. To predict the next word's token, we use the `predict` method and apply `np.argmax` to select the most likely next word. The predicted token is then mapped back to its corresponding word, which is appended to the existing seed text. The process is repeated until we achieve the desired number of words (`next_words`), and the function returns the fully generated text (`seed_text + next_words`)

9.   Let's generate the text:

```
input_text= "In the hustle and bustle of ipoti"
print(generate_text(input_text, 50, model,
    max_sequence_len))
```

We define the seed text, which is the `input_text` variable here. Our choice of next words we want the model to generate is `50`, and we pass in our trained model and also `max_sequence_len`. When we run the code, it returns the following output:

```
In the hustle and bustle of ipoti the city the friends also
learned about the wider context of the ancient world including
the people who had lived and worshipped in the area they
explored nearby archaeological sites and museums uncovering
artifacts and stories that shed light on the lives and beliefs
of those who had come before
```

The sample of generated text indeed resembles something you might find in a children's storybook. It continues the initial prompt coherently and creatively. While this example showcases the power of LSTMs as text generators, in this era, we leverage **large language models** (**LLMs**) for such applications. These models are trained on massive datasets, and they have a more sophisticated understanding of language; hence, they can generate more compelling stories when we prompt or fine-tune them properly.

We are now at the end of this chapter on NLP. You should now have the requisite foundational ideas to build your own NLP projects with TensorFlow. All you have learned here will also help you effectively navigate the NLP section of the TensorFlow Developer certificate exam. You have come a long way, and you should give yourself credit. We have one more chapter to go. Before we proceed to the chapter on time series, let's summarize what we learned in this chapter.

Summary

In this chapter, we embarked on a voyage through the world of RNNs. We began by looking at the anatomy of RNNs and their variants, and we explored the task of classifying news articles using different model architectures. We took a step further by applying pretrained word embeddings to our best-performing model in our quest to improve it. Here, we learned how to apply pretrained word embeddings in our workflow. For our final challenge, we took on the task of building a text generator to generate children's stories.

In the next chapter, we will examine time series, explore its unique characteristics, and uncover various methods of building forecasting models. We will tackle a time-series problem, where we will master how to prepare, train, and evaluate time-series data.

Questions

Let's test what we learned in this chapter:

1. Load the IMDB movies review dataset from TensorFlow and preprocess it.
2. Build a CNN movie classifier.
3. Build an LSTM movie classifier.
4. Use the GloVe 6B embedding to apply a pretrained word embedding to LSTM architecture.
5. Evaluate all the models, and save your best-performing one.

Further reading

To learn more, you can check out the following resources:

- Cho, K., et al. (2014). *Learning Phrase Representations using RNN Encoder–Decoder for Statistical Machine Translation.* arXiv preprint arXiv:1406.1078.

- Hochreiter, S., & Schmidhuber, J. (1997). *Long short-term memory.* Neural computation, 9(8), 1735–1780.

- Mikolov, T., Chen, K., Corrado, G., & Dean, J. (2013). *Efficient Estimation of Word Representations in Vector Space.* arXiv preprint arXiv:1301.3781.

- *The Unreasonable Effectiveness of Recurrent Neural Networks*: http://karpathy.github.io/2015/05/21/rnn-effectiveness/.

Part 4 –
Time Series
with TensorFlow

In this part, you will learn to build time series forecasting applications with TensorFlow. You will understand how to perform preprocess and build models for time series data. In this part, you will also learn to generate forecast for time series.

This section comprises the following chapters:

- *Chapter 12, Introduction to Time Series, Sequences, and Predictions*
- *Chapter 13, Time Series, Sequence and Prediction with TensorFlow*

12

Introduction to Time Series, Sequences, and Predictions

Time series cut across various industries, sectors, and aspects of our lives. Finance, healthcare, social sciences, physics – you name it, time series data is there. It's in sensors monitoring our environment, social media platforms tracking our digital footprint, online transactions recording our financial behavior, and many more avenues. This sequential data represents dynamic processes that evolve over time, and as we increasingly digitize our planet, the volume, and thereby the importance, of this data type is set to grow exponentially.

Time series follow a chronological order, capturing events as they occur in time. This temporal nature of time series bestows a unique quality that differentiates it from cross-sectional data. When we turn on the searchlight on time series data, we can observe attributes such as trends, seasonality, noise, cyclicity, and autocorrelations. These unique characteristics endow time series data with rich information, but they also present us with a unique set of challenges that we must overcome to harness the gains inherent in this data type. With frameworks such as **TensorFlow**, we can leverage patterns from the past to make informed decisions about the future.

In this chapter, we will be covering the following topics:

- Time series analysis – characteristics, applications, and forecasting techniques
- Statistical techniques for forecasting time series
- Preparing data for forecasting with neural networks
- Sales forecasting with neural networks

By the end of this chapter, you will have gained theoretical insight and hands-on experience in building, training, and evaluating time series forecasting models using statistical and deep learning techniques.

Time series analysis – characteristics, applications, and forecasting techniques

We know time series data is defined by the ordering of data points in a sequence over time. Imagine we are forecasting energy consumption patterns in London. Over the years, there has been a growing increase in energy consumption, perhaps due to urbanization – this signifies a positive upward trend. During winter each year, we expect energy consumption to rise as more people will need to heat up their homes and offices to stay warm. This seasonal change in the weather also accounts for seasonality in energy utilization. Again, we could also witness an unusual surge in energy consumption due to a major sporting event, leading to a large influx of guests during the period. This causes noise in the data as such events are one-offs or occur at irregular intervals.

In the following sections, let us explore the characteristics of time series, types, applications, and techniques for modeling time series data.

Characteristics of time series

To effectively build efficient forecasting models, we need to gain a clear understanding of the underlying nature of time series data. We may find ourselves working with time series data that has a positive upward trend, monthly seasonality, noise, and autocorrelation, while the next time series we work on may have yearly seasonality and noise but no visible sign of autocorrelation or trends in the data.

Understanding these data properties of time series data arms us with the requisite details to make informed preprocessing decisions. For example, if we are working with a dataset with high volatility, our knowledge of this may inform our decision to apply smoothing during our preprocessing steps. Later in this chapter, where we will be building statistical and deep learning models for forecasting time series, we will see how our understanding of the properties of time series data will guide our decisions with respect to engineering new features, choosing optimal hyperparameter values, and making model selection decisions. Let us examine the characteristics of time series:

- **Trend**: Trends refer to the general direction in which a time series is moving over the long term. We could look at a trend as the overall big picture of our time series data. Trends can be linear, as shown in *Figure 12.1*, or nonlinear (quadratic and exponential); they can also be in a positive (upward) or negative (downward) direction:

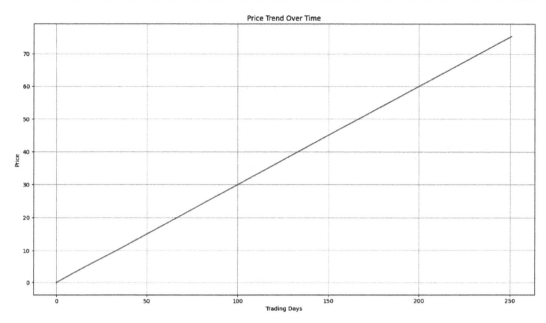

Figure 12.1 – A plot showing a positive stock price trend over time

Trend analysis empowers data professionals, businesses, and policymakers to make informed decisions about the future.

- **Seasonality**: Seasonality refers to repetitive cycles occurring at regular intervals over a specific period, such as on a daily, weekly, monthly, or yearly basis. These variations are often byproducts of seasonal fluctuations; for example, a retail store in a residential area might get higher sales on weekends compared to weekdays (weekly seasonality). The same store could also witness a surge in sales during the holiday season in December and a drop in sales shortly after the festive period (annual seasonality), as illustrated by *Figure 12.2*:

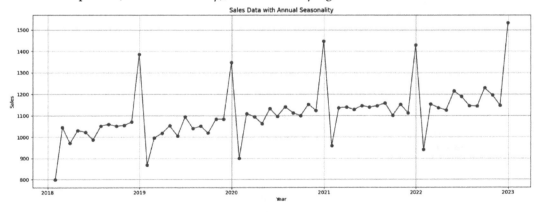

Figure 12.2 – A plot showing annual seasonality for a retail store

- **Cyclicality**: Cyclicality refers to irregular cycles that occur in a time series over a long period of time. Unlike seasonality, these cycles are long-term in nature and their duration and magnitude are irregular, making them harder to predict when compared to seasonality. Economic cycles are a good example of cyclicity. These cycles are influenced by several factors such as inflation, interest rates, and government policies. Due to its irregular nature, predicting the timing, duration, and magnitude of cycles can be quite challenging. Advanced statistical and machine learning models are often required to model and forecast cyclic patterns accurately.

- **Autocorrelation**: Autocorrelation is a statistical concept that refers to the correlation between a time series and the lagged version of itself, as shown in *Figure 12.3*.

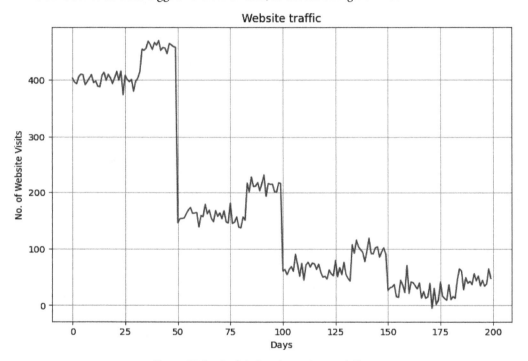

Figure 12.3 – A plot showing autocorrelation

It is often referred to as serial correlation and measures the degree to which a data point is related to its past values. Autocorrelations can be positive or negative, with values ranging from -1 to 1.

- **Noise**: Noise is an inherent part of any real-world data. It refers to the random fluctuations in data that cannot be explained by the model, nor can it be explained by any known underlying factors, patterns, or structural influences. These fluctuations can result from various sources such as measurement errors or unexpected events. For example, in the financial markets, an unpredicted event such as a political announcement can create noise that deviates stock prices from their underlying trends. Noise displays randomness and unexplained variations that we may have to smooth out when forecasting time series data.

We have discussed some important characteristics that can occur individually or together in time series data. Next, let us look at the types of time series data.

Types of time series data

Time series can be classified as the following:

- Stationary and non-stationary
- Univariate and multivariate

Stationary and non-stationary time series

A stationary time series is a time series whose statistical properties (mean, variance, and autocorrelation) remain constant over time. It is a series that displays recurring patterns and behaviors that are likely to replicate themselves in the future. A non-stationary time series is the opposite. It is not stationary, and we typically find these types of time series in many real-world scenarios where the series may display trends or seasonality. For example, we will expect the monthly sales of a ski resort to reach their peak during winter and dip during off-seasons. This seasonal component has an impact on the statistical properties of the series.

Univariate and multivariate time series

A univariate time series is a type of time series where we track just one metric over time. For example, we could use a smartwatch to track the number of steps we take on a daily basis. On the other hand, when we track more than one metric over time, we have a multivariate time series, as shown in *Figure 12.4*. In the chart, we see the interaction between inflation and wage growth in the UK. Over time, we see that inflation persistently outpaces wage growth, leaving everyday people with lowered real income and purchasing power:

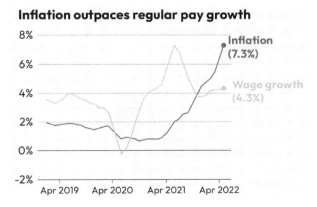

Figure 12.4 – A multivariate time series showing the relationship between inflation
and wage growth (https://www.bbc.co.uk/news/business-62218706)

Multivariate time series analysis lets us account for the dependencies and interactions between several variables over time. Next, let us delve into the importance of time series data and various applications of time series.

Applications of time series

We have discussed the types and properties of time series data. By applying machine learning techniques, we can leverage the wealth of information in these data types. Let us examine some of the important applications of machine learning in time series:

- **Forecasting**: We can apply machine learning models to forecast time series; for instance, we may want to forecast the future sales of a retail store to inform our inventory decisions. If we analyze the sales record of the business, we may find some patterns, such as increased sales during holiday seasons or lowered sales during specific months. We can train our models with these patterns to make informed predictions about the future sales of the store, allowing key stakeholders to effectively plan for the expected demand.

- **Imputed data**: Missing values can pose a significant challenge when we are working on analyzing or forecasting time series data. A good solution is the application of imputation, allowing us to fill the missing data points with substitute values. For example, in *Figure 12.5 (a)*, we see a plot of temperature values over a year. We quickly notice that some temperature recordings are missing. With the aid of imputation, we can estimate those values using adjacent days, as shown in *Figure 12.5 (b)*:

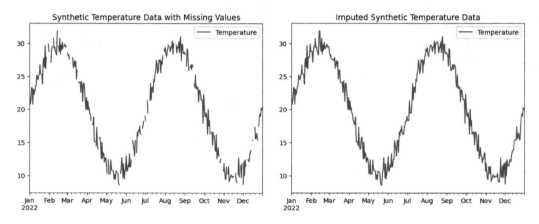

Figure 12.5 – A plot displaying temperature over time (a) with missing values (b) with no missing values

By filling in the gaps, we now have a complete dataset that can be better utilized for analysis and prediction.

- **Anomaly detection**: Anomalies are data points that deviate significantly from the general norm. We can apply time series analysis to detect anomalies and potentially identify significant issues. For example, in credit card transactions, as illustrated in *Figure 12.6*, an unusually large transaction might indicate fraudulent activity:

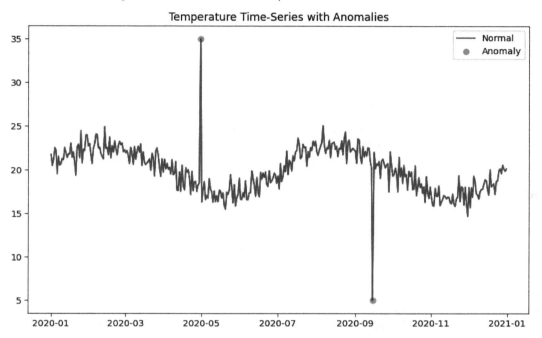

Figure 12.6 – A plot showing spikes in transaction values

By using time series analysis to identify such anomalies, the bank can swiftly take action to mitigate potential damages.

- **Trend analysis**: Understanding trends can provide valuable insights into the underlying phenomena. For example, the international energy agencies show that 14% of all new cars sold in 2022 were electric vehicles. This trend could indicate that people are moving toward a more sustainable option of transportation (see https://www.iea.org/reports/global-ev-outlook-2023/executive-summary).

- **Seasonality analysis**: Another useful application of time series is in seasonality analysis. This could prove useful in guiding energy consumption planning and infrastructural expansion needs.

We have now looked at some important applications of time series data. Next, let us take a look at some important techniques for forecasting time series.

Techniques for forecasting time series

In this book, we will examine two main techniques for forecasting time series: statistical methods and machine learning methods. Statistical methods use mathematical models to capture the trend, seasonality, and other components of the time series data, with popular models being **autoregressive integrated moving average (ARIMA)** and **seasonal and trend decomposition using LOESS (STL)**. However, these methods are beyond the scope of this book. Here, we will be using simpler statistical methods such as **naïve forecasting** and **moving averages** to establish our baseline, after which we will apply different machine learning methods. In this chapter, we will focus on using **deep neural networks (DNNs)**, and in the next chapter, we will apply **recurrent neural networks (RNNs)** and **long short-term memory networks (LSTMs)**.

Each approach has its pros and cons, and the best approach to forecasting time series is largely dependent on the specific characteristics of the data and the problem at hand. It is important thatwe highlight here that time series forecasting is a broad field, and there are other methods that fall outside the scope of this book that you may explore at a later stage. Before we move into modeling time series problems, let us examine how we can evaluate this type of data.

Evaluating time series forecasting techniques

To effectively evaluate a time series forecasting model, we must gauge its performance with appropriate metrics. In *Chapter 3*, Linear Regression with TensorFlow, we explored several regression metrics such as MAE, MSE, RMSE, and MAPE. We can apply these metrics to evaluate time series forecasting models. However, in this chapter, we will concentrate on the application of MAE and MSE in line with the exam requirements. We use MAE to compute the average of the absolute difference between the predicted and true values. This way, we have a sense of how wrong our predictions are. A smaller MAE indicates a better model fit. Imagine you are a stock market analyst trying to forecast the future price of a specific stock. Using MAE as your evaluation metric, you would get a clear understanding of how much, on average, your forecasts differ from the actual stock prices. This information can help you refine your model to make more accurate predictions, minimizing potential financial risks.

On the other hand, MSE takes the average of squared discrepancies between predictions and actuals. By squaring the errors, MSE is more sensitive to large errors compared to MAE, making it useful where large discrepancies are unfavorable, such as when working with a power grid where precise load forecasting is of top priority. With this in mind, let's now turn our attention to a sales use case and apply our learnings to forecast future sales.

Retail store forecasting

Imagine you are working as a machine learning engineer and your company just landed a new project. A rapidly growing superstore in Florida wants your help. They want to predict future reviews, as this will serve as a guide in planning the expansion of their stores to meet the expected demand. You have been saddled with the responsibility of building a forecasting model with the available historical data provided by the Tensor superstore. Let's jump in and see how you can solve this problem, as your company is counting on you. Let's get started!

1. We begin by importing the necessary libraries for our project:

```
import pandas as pd
import numpy as np
import matplotlib.pyplot as plt
```

Here, we import numpy and matplotlib for numerical analysis and visualization purposes and pandas for data transformations.

2. Next, we load the time series data:

```
df = pd.read_csv('/content/sales_data.csv')
df.head()
```

Here, we load the data and use the head function to get a snapshot of the first five rows of the data. When we run the code, we see that the first day of sales from the data given to us is 2013-01-01. Next, let us look at some statistics to get a sense of the data we have in hand.

3. Use the following code to check the data type and summary statistics:

```
# Check data types
print(df.dtypes)

# Summary statistics
print(df.describe())
```

When we run the code, it returns the data type as float64 and key summary statistics for our sales data. Using the describe function, we get a count of 3,653 data points. This points to daily data over a 10-year period. We also see the mean sales per day come to around $75, giving us a sense of the central tendency. We see decent variability in the daily sales with a standard deviation of 20.2. The minimum and maximum values reveal a range from $22 to $128 in sales, signaling some significant fluctuations occurring. The 25th and 75th percentiles are 60.27 and 89.18, respectively, showing that lower-volume days see sales around $60 while higher-volume days see around $90. Let us continue to explore the data by looking at it on a plot.

4. Let's visualize our data:

```
#Sales data plot
df.set_index('Date').plot()
plt.ylabel('Sales')
plt.title('Sales Over Time')
plt.xticks(rotation=90)
plt.show()
```

The code returns the plot shown in *Figure 12.7*, which represents the company's sales over a 10-year period:

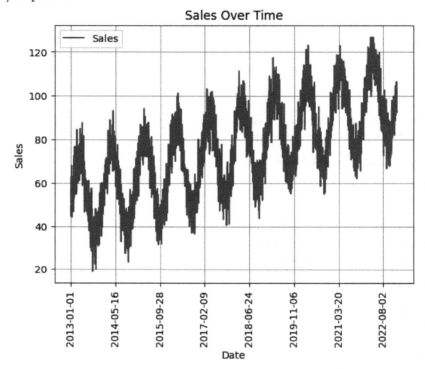

Figure 12.7 – A plot showing sales over time

From the plot in *Figure 12.7*, we can observe an overall positive upward trend, potentially indicative of economic growth or successful business strategies, such as new product releases and effective marketing. A clear yearly seasonality also emerges; this may suggest that the company deals in seasonal goods with annual demand fluctuations. Also, we observe some noise exists. This could be a result of weather variability, random events, or the entry of competitors. The upward trend in the data demonstrates promising performance and growth in sales over time. However, the seasonal effects and noise elements showcase complex dynamics underneath the aggregate trend. Next, let's explore how to data partition the data.

Data partitioning

In time series forecasting, we typically divide our dataset into distinct sections: a training period for training our machine learning models, a validation period for model tuning and evaluation, and a test period for assessing performance on unseen data. This process is known as fixed partitioning. An alternate method is roll-forward partitioning, which we will be discussing shortly.

Our sales data demonstrates seasonality; hence, we have to split the data in such a way that each partition captures entire seasonal cycles. We do this to ensure we do not omit important seasonal patterns in one or more of our partitions. While this method diverges from typical data partitioning, we see that when working with other machine learning problems where random samples are taken to form training, validation, and testing sets, the fundamental purpose remains the same. We train our model on the training data, fine-tune it using the validation data, and evaluate it on the test data. We can then incorporate the validation data into the training data to benefit from the most recent information and forecast future data.

It is generally a good idea to ensure that your training set is large enough to capture all relevant patterns in the data, including any seasonal behavior of the data. When setting the size of your validation set, you must strike a balance. While a larger validation set gives a more reliable estimate of model performance, it also reduces the size of the training set. You should also remember to retrain your final model on the entire dataset (combining the training and validation sets) before making final predictions. This strategy maximizes the amount of data the model learns from, likely improving its predictive performance on future unseen data. Again, avoid shuffling the data before splitting, as it would disrupt the temporal order, leading to misleading results. In fixed partitioning, we usually use a chronological split such that the training data should be from the earliest timestamps, followed sequentially by the validation set, and finally, the test set containing the latest timestamps. Let's split our sales data into training and validation sets:

```
# Splitting the data into training and validation sets
split_time = int(0.8 * len(df))
time = np.arange(len(df))
x_train = df['Sales'][:split_time]
x_valid = df['Sales'][split_time:]
time_valid = time[split_time:]
```

Here, we split the data into training and validation sets. We take 80% of the data (`len(df) * 0.8`), which, in this case, is 8 years of data for training and the last 2 years for validation. We use the `int` function to ensure that the split time is an integer for indexing purposes. We set up our training and validation sets, using everything before the split time for training and everything after the split time for validation.

Next, let us plot our training and validation data:

```
fig, ax = plt.subplots(figsize=(12, 9))
# Plot training and validation data
ax.plot(df.index[:split_time], x_train, label='Training data')
ax.plot(df.index[split_time:], x_valid, label='Validation data')
# Add vertical split line
ax.vlines(df.index[split_time], ymin=0, ymax=df['Sales'].max(),
colors='r', linestyles='--')
ax.set_ylabel('Sales')
```

```
ax.set_title('Fixed Partitioning')
ax.legend()
plt.grid(True)
plt.tight_layout()
plt.show()
```

This code displays the partitioning of sales data into a training set and a validation set, marking them in green and blue colors respectively, with dates along the *x* axis and sales along the *y* axis. For readability, we use the dotted lines to show the split between our training set and validation set:

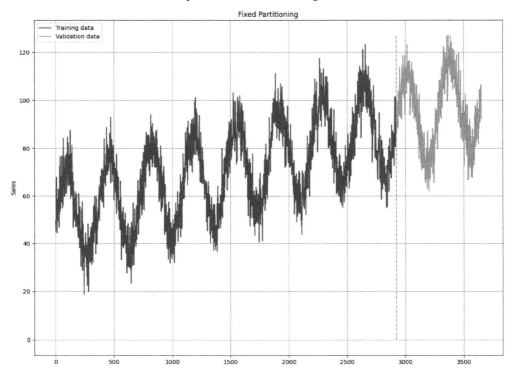

Figure 12.8 – A plot showing fixed partitioning

In *Figure 12.8*, we split our sales data into 8 years for training and 2 years for validation. In this scenario, our test set will be data from the future. This is done to ensure that the model is trained on the earliest part of our series, validated on the recent past, and tested on the future. Another method of partitioning time series data is called roll-forward partitioning, or "walk-forward" validation. In this method, we start with a short training period and gradually increase it. For each training period, the following period is used as the validation set. This mirrors a real-life situation where we continually retrain our model as new data comes in and use it to predict the next period. Let's discuss our first method of forecasting, called naïve forecasting.

Naïve forecasting

Naïve forecasting is one of the simplest methods for forecasting in time series analysis. The principle behind naïve forecasting is to simply set all forecasts to be the value of the last observed point. This is why it's referred to as "naïve." It is a method that assumes that the future value is likely to be the same as the current one. Despite its simplicity, naïve forecasting can often serve as a good baseline for time series forecasting; however, its performance can vary depending on the characteristics of the time series.

Let us see how to implement this in code:

1. Let's implement naïve forecasting:

    ```
    # Apply naive forecast
    df['Naive_Forecast'] = df['Sales'].shift(1)
    df.head()
    ```

    To implement the naïve forecasting method, each forecasted value is simply set to the actual observed value from the previous time step, achieved by shifting the `Sales` column by one unit. We use `df.head()` to display the first five rows of the DataFrame, providing a quick overview of the sales data and the naïve forecast:

	Date	Sales	Naive_Forecast
0	2013-01-01	48.40	NaN
1	2013-01-02	49.63	48.40
2	2013-01-03	44.66	49.63
3	2013-01-04	56.70	44.66
4	2013-01-05	62.87	56.70

    Figure 12.9 – A snapshot of the DataFrame showing the sales and naïve forecasts

    From the table in *Figure 12.9*, we see that the first forecast is not available. This is because this method simply takes all the values in the series starting from one step before the validation data until the second-last value of the series. This effectively shifts the time series by one time step into the future.

2. Let us create a set of functions for plotting and evaluating purposes:

    ```
    def plot_forecast(time, x_valid, forecast, title, start_
    index=None, end_index=None):
        plt.figure(figsize=(10, 6))
        plt.plot(time, x_valid, label="True Values")
        plt.plot(time, forecast, label=title)

        if start_index and end_index:
            tick_frequency = (end_index - start_index) // 10
            tick_indices = list(range(start_index, end_index, tick_
    frequency))
    ```

```
            tick_labels = [df['Date'].iloc[i] for i in tick_indices]
            plt.xticks(tick_indices, tick_labels, rotation=90)
            plt.xlim(start_index, end_index)

    plt.title(f"Validation Split: True Values vs {title}")
    plt.legend()
    plt.xlabel("Time")
    plt.ylabel("Sales")
    plt.grid(True)
    plt.tight_layout()
    plt.show()

def calculate_errors(y_true, y_pred):
    mse = tf.keras.metrics.mean_squared_error(y_true, y_pred).
numpy()
    mae = tf.keras.metrics.mean_absolute_error(y_true, y_pred).
numpy()
    print('Mean Squared Error:', mse)
    print('Mean Absolute Error:', mae)
    return mse, mae
plot_forecast(time_valid, x_valid, df['Naive_Forecast'][split_
time:], "Naive Forecast")
```

We pass in the required parameters into the plot_forecast function and run the code to generate the following plot:

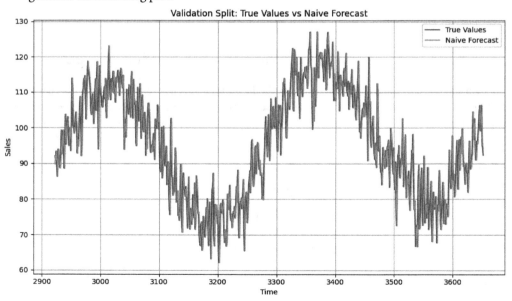

Figure 12.10 – A time series forecast using the naïve method

The plot in *Figure 12.10* displays the forecasted values and the true values for the validation data. Because the plot looks a bit clustered due to the closeness in values, let us zoom in to help us visually investigate how our forecasting is doing.

3. Let's look at a specific time range:

```
# Zoomed in Plot
start_date = "2022-01-01"
end_date = "2022-06-30"
start_index = df[df["Date"] == start_date].index[0]
end_index = df[df["Date"] == end_date].index[0]
plot_forecast(time_valid, x_valid, df['Naive_Forecast'][split_
time:], "Naive Forecast", start_index, end_index)
```

We set the start date and end date parameters to 2022-01-01 and 2022-06-30, respectively. The resulting plot is as follows:

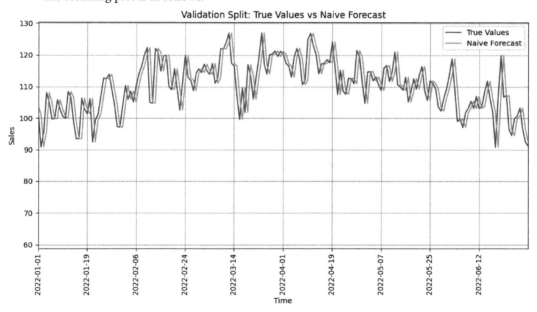

Figure 12.11 – Zoomed-in time series forecasting using the naïve method

From the plot in *Figure 12.11*, we can see that it starts the forecast one step later because the naïve forecast is shifted one step into the future.

4. Next, let us evaluate the performance of the naïve method using the `calculate_errors` function:

```
calculate_errors(x_valid, df['Naive_Forecast'][split_time:])
```

The `calculate_errors` function returns the MSE and MAE values. When we run the code, we achieve an MAE and MSE values of `4.87` and `38.19` respectively. Recall that for the MSE and MAE values, lower values are always better.

Let's keep this in mind as we explore other forecasting techniques. Naïve forecasting can serve as a baseline to compare the performance of more complex models. Next, let's examine another statistical method, called moving average.

Moving average

Moving average is a technique for smoothing time series data by replacing each point with an average of the neighboring points. In this process, we generate a new series in which the data points are averages of the raw data in our original series. The key parameter in this method is the **window width**; this determines the number of consecutive raw data points included in the average calculation.

The term "moving" refers to the sliding of the window along the time series to compute average values, thereby generating a new series.

Types of moving averages

Two main types of moving averages are commonly used in time series analysis – the centered moving average and the trailing moving average:

- **Centered moving average**: A centered moving average calculates the average around a central point (*t*). It uses data from both before and after the time of interest for visualization, as illustrated in *Figure 12.12*. Centered moving averages can give a well-balanced view of data trends, but since they require future data, they're not suitable for forecasting as we do not have access to future values when making forecasts. Centered moving averages are good for visualization and time series analysis.

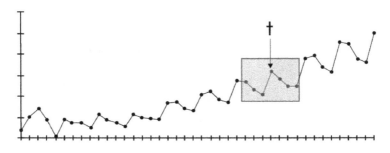

Figure 12.12 – A plot showing a centered moving average

Centered moving averages can give a well-balanced view of data trends, but since they require future data, they're not suitable for forecasting as we do not have access to future values when making forecasts. Centered moving averages are good for visualization and time series analysis.

- **Trailing moving average**: A trailing moving average, also known as a rolling or running average, calculates the average using the most recent *n* data points. This method solely requires past data points, as illustrated in *Figure 12.13*, making it ideal for forecasting.

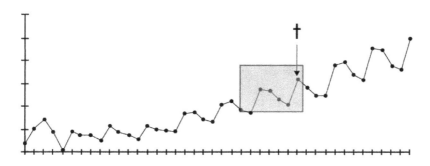

Trailing moving average

window from time *t* and backward

Figure 12.13 – A plot showing a trailing moving average

To compute a trailing moving average, the first step is choosing the window width (W). This selection can depend on various factors such as the series' patterns and how much smoothing you want to achieve. A smaller window width will track quick changes closely, but this will happen at the risk of including more noise. On the other hand, a larger window width will provide a smoother line but might miss out on some short-term fluctuations.

Let's see how to implement moving averages:

```
def moving_average_forecast(df, window):
    return df['Sales'].rolling(window=window).mean().shift(1)
# Moving Average Forecast for 30 days
df['Moving_Average_Forecast'] = moving_average_forecast(df, 30)
```

We set up a function called moving_average_forecast. It uses pandas' rolling function to calculate the moving average of the Sales data over a specified window, then shifts the outcome one step forward to simulate a forecast for the subsequent time step, when we apply this function with a window size of 30 days it returns to us the result in a new Moving_Average_Forecast column. You can think of it as using a window of 30 days of sales to predict the sales on the 31st day.

We call the `plot_forecast` function to plot both the validation data and the moving average forecast data. We can see the resulting plot in *Figure 12.14*:

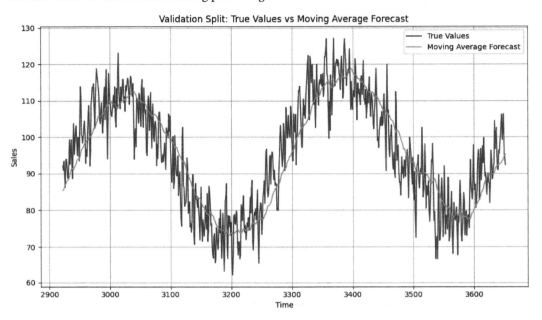

Figure 12.14 – A time series forecast using the moving average method

In the plot in *Figure 12.14*, the moving average forecast is computed using the past 30 days of sales data. A smaller window size, such as a 7-day window, will follow the actual sales more closely than the 30-day moving average, but it might also capture more of the noise in the data.

The next logical step will be to evaluate our model by using the `calculate_errors` function again. This time, we pass in the moving average forecast against the true validation values:

```
# Compute and print mae and mse
calculate_errors(x_valid, df['Moving_Average_Forecast'][split_time:])
```

When we run the code, we achieve an MSE of 50.40 and an MAE of 5.70 for the moving average forecast. The MAE and MSE are much worse than our baseline. If you change the window size to 7 days, we end up with a much lower MSE and MAE of 39.80 and 5.038 respectively. This is a much better result, but still worse than our naïve approach. You could experiment with a smaller window size and see whether your results will surpass the baseline.

However, we need to note that the underlying assumption when using moving average is stationarity. And we know that our time series has both trend and seasonality, so how do we achieve stationarity with this data? And will this help us to achieve a much lower MAE? To achieve stationarity, we use a concept called **differencing**, however, this is outside the scope of this book. Next, let us see how we can perform forecasting with machine learning, and in particular, with neural networks with TensorFlow.

Time series forecasting with machine learning

So far, we have examined statistical methods with reasonable success. Now, we will proceed with modeling time series data using deep learning techniques. We will begin with mastering how to set up a window dataset. We will also cover ideas such as shuffling and batching, and see how we can build and train a neural network for our sales forecasting problem. Let's begin by mastering how we can prepare time series data for modeling using the windowed dataset method with the aid of TensorFlow utilities:

1. We begin by importing the libraries required:

    ```
    import tensorflow as tf
    import numpy as np
    ```

    Here, we will be using NumPy and TensorFlow to prepare and manipulate our data into the required structure for modeling.

2. Let us create a simple dataset. Here, we are assuming the data consists of temperature values for two weeks:

    ```
    # Create an array of temperatures for 2 weeks
    temperature = np.arange(1, 15)
    print(temperature)
    ```

    When we print out the temperature, we get the following:

    ```
    [ 1  2  3  4  5  6  7  8  9 10 11 12 13 14]
    ```

    We get an array of values 1-14, where we are assuming the temperature rises from 1 on the first day to 14 on the 14th day. Odd, but let's assume this is the case.

3. Let's create windowed data. Now that we have our data, we need to make a "window" of data points:

    ```
    window_size = 3
    batch_size = 2
    shuffle_buffer = 10
    ```

    The `window_size` parameter refers to the window of data under consideration. If we set the window size to 3, this means we will use 3 consecutive days' temperature values to predict the next day's temperature. The batch size determines how many samples are processed in each iteration during training, and `shuffle_buffer` specifies the number of elements from which TensorFlow randomly samples when shuffling the data. We shuffle to avoid sequential bias; we will expand on this in the next chapter.

4. Creating a dataset works as follows:

    ```
    dataset = tf.data.Dataset.from_tensor_slices(temperature)
    for element in dataset:
        print(element.numpy())
    ```

This line of code is used to create a TensorFlow `Dataset` object from our temperature data. This `Dataset` API is a high-level TensorFlow API for reading data and transforming it into a form that a machine learning model can use. Next, we iterate over the dataset and print each element. We use `numpy()` to convert a TensorFlow object into a NumPy array. When we run the code, we get the numbers 1-14; however, they are now ready to be windowed.

5. Next, let's transform our temperature data into a "windowed" dataset:

```
dataset = dataset.window(window_size + 1, shift=1, drop_
remainder=True)
for window in dataset:
    window_data = ' '.join([str(
        element.numpy()) for element in window])
    print(window_data)
```

We apply the `window` method to create a dataset of windows, where each window is a dataset itself. The `window_size + 1` parameter means that we are considering the `window_size` elements as input and the next one as a label. The `shift=1` parameter means that the window moves one step at a time. The `drop_remainder=True` parameter means that we will drop the last few elements if they can't form a complete window.

When we print out our `window_data`, we get the following output:

```
After window:
1 2 3 4
2 3 4 5
3 4 5 6
4 5 6 7
5 6 7 8
6 7 8 9
7 8 9 10
8 9 10 11
9 10 11 12
10 11 12 13
11 12 13 14
```

We see that we now have the window size that we set to 3 values and the (+1) that will serve as our label. Because we set the shift value to 1, the next window starts from the second value in our series, which in this case is 2. Next, the window will move one more step until we make all the windowed data.

6. Flattening the data works as follows:

```
dataset = dataset.flat_map(
    lambda window: window.batch(window_size + 1))
for element in dataset:
    print(element.numpy())
```

It's easy to view each window created in the last step as a separate dataset. With this code, we flatten the data so that each window of data is packaged as a single batch in the main dataset. We use `flat_map` to flatten it back into a dataset of tensors and `window.batch(window_size + 1)` to convert each window dataset into a batched tensor. When we run the code, we get the following:

```
After flat_map:
[1 2 3 4]
[2 3 4 5]
[3 4 5 6]
[4 5 6 7]
[5 6 7 8]
[6 7 8 9]
[ 7  8  9 10]
[ 8  9 10 11]
[ 9 10 11 12]
[10 11 12 13]
[11 12 13 14]
```

We can see the windowed data is now put into batched tensors.

7. Shuffle the data as follows:

```
dataset = dataset.shuffle(shuffle_buffer)
print("\nAfter shuffle:")
for element in dataset:
    print(element.numpy())
```

In this code block, we shuffle the data. This is an important step as shuffling is used to ensure that the model doesn't accidentally learn patterns from the order in which examples are presented during training. When we run the code block, we get the following:

```
After shuffle:
[5 6 7 8]
[4 5 6 7]
[1 2 3 4]
[11 12 13 14]
[ 7  8  9 10]
[ 8  9 10 11]
[10 11 12 13]
[ 9 10 11 12]
[6 7 8 9]
[2 3 4 5]
[3 4 5 6]
```

We can see that the mini datasets in the main dataset have been shuffled. However, take note that the features (window) and the label are unchanged in the shuffled dataset.

8. Mapping features and labels works as follows:

```
dataset = dataset.map(lambda window: (window[:-1],
    window[-1]))
print("\nAfter map:")
for x,y in dataset:
    print("x =", x.numpy(), "y =", y.numpy())
```

The map method applies a function to each element of the dataset. Here, we are splitting each window into features and labels. The features are all but the last element of the (window[:-1]) window, and the label is the last element of the (window[-1]) window. When we run the code block, we see the following:

```
After map:
features = [3 4 5] label = 6
features = [1 2 3] label = 4
features = [10 11 12] label = 13
features = [ 8  9 10] label = 11
features = [4 5 6] label = 7
features = [7 8 9] label = 10
features = [11 12 13] label = 14
features = [5 6 7] label = 8
features = [2 3 4] label = 5
features = [6 7 8] label = 9
features = [ 9 10 11] label = 12
```

From our print result, we see the features are made up of three observations in an order and the next value is our label.

9. Batching and prefetching the data works as follows:

```
dataset = dataset.batch(batch_size).prefetch(1)
print("\nAfter batch and prefetch:")
for batch in dataset:
    print(batch)
```

The batch() function groups the dataset into batches of batch_size. In this case, we're creating batches of size 2. The prefetch(1) performance optimization function makes sure that TensorFlow always has one batch ready to go while it's processing the current one. After these transformations, the dataset is ready to be used for training a machine learning model. Let's print out the batch:

```
After batch and prefetch:
(<tf.Tensor: shape=(2, 3), dtype=int64,
    numpy=array([[10, 11, 12],
```

```
           [ 3,   4,   5]])>, <tf.Tensor: shape=(2,),
           dtype=int64, numpy=array([13,   6])>)
(<tf.Tensor: shape=(2, 3), dtype=int64,
    numpy=array([[ 9, 10, 11],
          [11, 12, 13]])>, <tf.Tensor: shape=(2,),
           dtype=int64, numpy=array([12, 14])>)
(<tf.Tensor: shape=(2, 3), dtype=int64,
    numpy=array([[6, 7, 8],
          [7, 8, 9]])>, <tf.Tensor: shape=(2,),
           dtype=int64, numpy=array([ 9, 10])>)
(<tf.Tensor: shape=(2, 3), dtype=int64,
    numpy=array([[1, 2, 3],
          [4, 5, 6]])>, <tf.Tensor: shape=(2,),
           dtype=int64, numpy=array([4, 7])>)
(<tf.Tensor: shape=(2, 3), dtype=int64,
    numpy=array([[2, 3, 4],
          [5, 6, 7]])>, <tf.Tensor: shape=(2,),
           dtype=int64, numpy=array([5, 8])>)
(<tf.Tensor: shape=(1, 3), dtype=int64, numpy=array(
    [[ 8,   9, 10]])>, <tf.Tensor: shape=(1,),
      dtype=int64, numpy=array([11])>)
```

We see that each element of the dataset is a batch of features and label pairs, where the features are arrays of the `window_size` values from the original series, and the label is the next value that we want to predict. We have seen how to prepare our time series data for modeling; we have sliced, windowed, batched, shuffled, and split it into features and labels for this purpose.

Next, let us use what we have learned here on our synthetic sales data to forecast future sales values.

Sales forecasting using neural networks

Let's return to the sales data that we created to forecast sales using both naïve and moving average methods. Let us now use a neural network; here, we will use a DNN:

1. Extracting the data works as follows:

    ```
    time = pd.to_datetime(df['Date'])
    sales = df['Sales'].values
    ```

    In this code, we extract the `Date` and `Sales` data from the `Sales` DataFrame. `Date` is converted into datetime format and `Sales` is converted into a NumPy array.

2. Splitting the data works as follows:

```
split_time = int(len(df) * 0.8)
time_train = time[:split_time]
x_train = sales[:split_time]
time_valid = time[split_time:]
x_valid = sales[split_time:]
```

For uniformity, we split data into a training set and a validation set using the same 80:20 split by using `split_time` of 80%.

3. Creating the windowed dataset works as follows:

```
def windowed_dataset(
    series, window_size, batch_size, shuffle_buffer):
        dataset = tf.data.Dataset.from_tensor_slices(
            series)
        dataset = dataset.window(window_size + 1,
            shift=1, drop_remainder=True)
        dataset = dataset.flat_map(lambda window:
            window.batch(window_size + 1))
        dataset = dataset.shuffle(shuffle_buffer).map(
            lambda window: (window[:-1], window[-1]))
        dataset = dataset.batch(
            batch_size).prefetch(1)
    return dataset
window_size = 7
batch_size = 32
shuffle_buffer_size = 1000
dataset = windowed_dataset(x_train, window_size,
    batch_size, shuffle_buffer_size)
```

We create the `windowed_dataset` function; this function takes in a series, a window size, a batch size, and a shuffle buffer. It creates windows of data for training, with each window containing a `window_size + 1` data point. These windows are then shuffled and mapped to features and labels, where the features are all data points in a window, except the last one, and the label is the last data point. The windows are then batched and prefetched for efficient data loading.

4. Building the model works as follows:

```
model = tf.keras.models.Sequential([
    tf.keras.layers.Dense(10,
        input_shape=[window_size], activation="relu"),
    tf.keras.layers.Dense(10, activation="relu"),
```

```
            tf.keras.layers.Dense(1)
            ])
    model.compile(loss="mse",
        optimizer=tf.keras.optimizers.SGD(
            learning_rate=1e-6, momentum=0.9))
```

Next, we use a simple **feedforward neural network (FFN)** for modeling. This model contains two dense layers with ReLU activation, followed by a dense output layer with a single neuron. The model is compiled with MSE loss and **stochastic gradient descent (SGD)** as the optimizer.

5. Training the model works as follows:

```
    model.fit(dataset, epochs=100, verbose=0)
```

The model is trained on the windowed dataset for 100 epochs.

6. Generating predictions works as follows:

```
    input_batches = [sales[time:time + window_size][
        np.newaxis] for time in range(len(
            sales) - window_size)]
    inputs = np.concatenate(input_batches, axis=0)
    forecast = model.predict(inputs)
    results = forecast[split_time-window_size:, 0]
```

Here, we generate predictions in batches to improve computational efficiency. Only the predictions for the validation period are kept.

7. Evaluating the model works as follows:

```
    print(tf.keras.metrics.mean_squared_error(x_valid,
        results).numpy())
    print(tf.keras.metrics.mean_absolute_error(x_valid,
        results).numpy())
```

The MSE and MAE between the true validation data and the predicted data are calculated. Here, we achieved an MSE of 34.51 and an MAE of 4.72, which surpasses all our simple statistical methods.

8. Visualizing the results works as follows. The true validation data and the predicted data are plotted over time to visualize the model's performance:

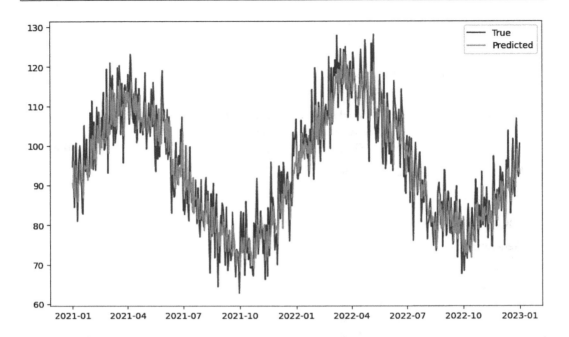

Figure 12.15 – A time series forecast using a simple FFN

From the plot in *Figure 12.15*, we see that our forecasted values map closely to the ground truth on the validation set, less the noisy spikes. Our FFN has demonstrated notable achievements in this experiment. Compared to the traditional statistical methods, our model exhibits a significant improvement in performance. You could tune the hyperparameters as well as implement callbacks to improve performance. However, our job is done here.

Let's meet in the final chapter, where we will be predicting Apple stock prices. See you there.

Summary

In this chapter, we explored the concept of time series, examined the core characteristics and types of time series, and looked at some well-known applications of time series in machine learning. We also covered concepts such as trailing and centered windows and examined how to prepare time series for modeling with neural networks with the aid of utilities from TensorFlow. In our case study, we applied both statistical and deep learning techniques in order to build a sales forecasting model for a fictional company.

In the next chapter, we will extend our modeling using more complex architectures such as RNNs, CNNs, and CNN-LSTM architecture in forecasting time series data. Also, we will explore concepts such as learning rate scheduler and Lambda layers. To conclude the final chapter of this book, we will build a forecasting model for Apple's closing stock price.

Questions

1. Apply the principle of naïve forecasting to the "*Air Passenger*" dataset using the exercise notebook provided.

2. Implement the moving average technique on the same dataset and evaluate its performance by calculating the MAE and MSE values.

3. Next, introduce the method of differencing to your moving average model. Again, assess the accuracy of your forecast by determining the MAE and MSE values.

4. With the sample temperature dataset at hand, demonstrate how to create meaningful features and labels from this data.

5. Lastly, experiment with the simple FFN model on the dataset and observe its performance.

13
Time Series, Sequences, and Prediction with TensorFlow

Welcome to the final chapter of our journey with TensorFlow. In the last chapter we examined how to build time series models using DNNs along with some simple statistical methods. Here, we will extend our learning by exploring advanced architectures such as CNNs, RNNs, LSTMs, CNN-LSTMs architectures. Also, we will see how to integrate in-built learning rate schedules from TensorFlow into our workflow to dynamically adjust the model's learning rate during our modeling process. We will also see how to apply custom learning rate schedulers to find the optimal learning rate when building models.

Next, we will discuss Lambda layers and how these arbitrary layers can be applied in our model architecture to enhance quick experimentation, enabling us to reshape our input seamlessly when working with more complex models such as LSTMs and RNNs. To conclude this chapter, we will extract Apple stock closing price data from Yahoo Finance and apply these models to build a forecasting model to predict future Apple stock prices.

In this chapter, we will be covering the following topics:

- Understanding and applying in-built learning rate schedules
- Using custom learning rate schedulers
- Utilizing Lambda layers in TensorFlow
- Employing RNNs, LSTMs, and CNNs for time series forecasting
- Apple stock price prediction using neural networks.

By the end of this chapter, you will have a deeper understanding of time series forecasting with TensorFlow, along with hands-on experience in applying different techniques in building time series forecasting models for real-world projects. Let's get started.

Understanding and applying learning rate schedules

In *Chapter 12*, Introduction to Time Series, Sequences, and Predictions, we built a DNN that achieved a **mean absolute error (MAE)** of 4.72. While this result was much better than our basic statistical methods, our next line of thought was how we could improve the performance of our DNN. One way of doing this is by finding the optimal learning rate. In *Chapter 7*, Image Classification with Convolutional Neural Networks, we discussed the important role of the learning rate in our modeling process as it controls the optimization process. Manually updating the learning rate can be a laborious process as the challenge lies in pinpointing what value works best. To have better control over the learning process, we apply a built-in learning rate schedule that adapts the learning rates based on defined criteria such as the number of epochs. Let us examine some built-in learning rate schedules:

- `ExponentialDecay`: This starts with a specified learning rate and decreases exponentially after a certain number of steps.

- `PiecewiseConstantDecay`: This provides a piecewise constant learning rate, where you specify boundaries and learning rates to divide the training process into several stages with different learning rates.

- `PolynomialDecay`: The `PolynomialDecay` schedule steadily reduces the learning rate of an optimizer from an initial learning rate to a specified end learning rate over a defined number of decay steps. A polynomial function is utilized to calculate the decayed learning rate at each step based on the initial and end values defined.

Let's examine how we can apply in-built learning rate schedule to the feedforward network we used in *Chapter 12, Introduction to Time Series, Sequences, and Predictions*. We are using the same sales data, but this time we will be applying different learning rate schedules in an attempt to improve the performance of the model. Let's begin:

1. We start by importing the libraries for this project:

```
import numpy as np
import pandas as pd
import matplotlib.pyplot as plt
from sklearn.preprocessing import MinMaxScaler
from sklearn.metrics import mean_squared_error, mean_absolute_
error
import tensorflow as tf
from tensorflow.keras.layers import Dense
from tensorflow.keras.models import Sequential
```

2. Next, let us load our dataset:

```
# Load data
url = 'https://raw.githubusercontent.com/PacktPublishing/
TensorFlow-Developer-Certificate-Guide/main/Chapter%2012/sales_
```

```
data.csv'
# Load the dataset
df = pd.read_csv(url)
```

We load the sales data from the GitHub repo for this book and put the CSV data into a DataFrame.

3. Now, we convert the `Date` column into datetime and make it the index:

```
df['Date'] = pd.to_datetime(df['Date'])
df.set_index('Date', inplace=True)
```

The first line of code converts the date column into a datetime format. We do this to easily perform time series operations. Next, we change the date column by setting it as the index of our DataFrame. This makes it easier to slice and dice our data using dates.

4. Let's extract the sales values from the DataFrame:

```
data = df['Sales'].values
```

Here, we are extracting the sales values from our sales DataFrame and converting them into a NumPy array. We will use this NumPy array to create our sliding window data.

5. Next, we'll create a sliding window:

```
window_size = 20
X, y = [], []
for i in range(window_size, len(data)):
    X.append(data[i-window_size:i])
    y.append(data[i])
```

Just as we did in *Chapter 12*, Introduction to Time Series, Sequences, and Predictions, we are using the sliding window technique to convert our time series into a supervised learning problem made up of features and labels. Here, the window size of 20 serves as our X feature, which contains 20 consecutive sales values, and our y is the next immediate value after those first 20 sales values. Here, we use the first 20 values to predict the next value.

6. Now, let's split our data into training and validation sets:

```
X = np.array(X)
y = np.array(y)
train_size = int(len(X) * 0.8)
X_train, X_val = X[:train_size], X[train_size:]
y_train, y_val = y[:train_size], y[train_size:]
```

We convert our data into NumPy array and split our data into training and validation sets. We use 80 percent of our data for training and 20 percent of our data for the validation set that we will use to evaluate our model.

7. Our next goal is to build out a TensorFlow dataset, which is a more efficient format for training models in TensorFlow:

```
batch_size = 128
buffer_size = 1000
train_data = tf.data.Dataset.from_tensor_slices(
    (X_train, y_train))
train_data = train_data.cache().shuffle(
    buffer_size).batch(batch_size).prefetch(
    tf.data.experimental.AUTOTUNE)
```

We apply the `from_tensor_slices()` method to make a dataset from the NumPy arrays. After this, we use the `cache()` method to speed up training by caching our dataset in memory. We apply the `shuffle(buffer_size)` method to randomly shuffle our training data to prevent issues such as `sequential bias`. Then we use the `batch(batch_size)` method to split our data into batches of a specified size; in this case, batches of 128 are fed into our model during training. Next, we use the `prefetch` method to ensure our GPU/CPU will always have data ready for processing, reducing the waiting time between the processing of one batch and the next. We pass in the `tf.data.experimental.AUTOTUNE` argument to tell TensorFlow to automatically determine the optimal number of batches to prefetch. This makes our training process smoother and faster.

8. Let's define our plotting and evaluation functions:

```
#evaluation function
def evaluate_model(model, X_val, y_val):
    forecast_val = model.predict(X_val)
    mae_val = mean_absolute_error(y_val, forecast_val)
    mse_val = mean_squared_error(y_val, forecast_val)
    return mae_val, mse_val

#plotting function
def plot_predictions(forecast_val, y_val, title="Baseline model
plot", start_date=None, end_date=None):
    plt.figure(figsize=(10, 6))

    plt.plot(forecast_val, label='Predicted Validation')
    plt.plot(y_val, label='Actual Validation')
    if start_date and end_date:
        start_idx = df.index.get_loc(pd.Timestamp(start_date))
        end_idx = df.index.get_loc(pd.Timestamp(end_date))
        start_idx -= len(data) - len(y_val)
        end_idx -= len(data) - len(y_val)
        plt.xlim(start_idx, end_idx)
```

```
plt.title(title)
plt.legend()
plt.show()
```

Our data is now ready for modeling. Let us explore using this data with in-built learning rate schedules from TensorFlow, after which we will look at how to find the optimal learning rate with a custom learning rate scheduler.

In-built learning rate schedules

We will be using the same model as we did in *Chapter 12, Introduction to Time Series, Sequences, and Predictions*. Let's define our model and explore the inbuilt learning rate schedules:

1. We'll start with the model definition:

    ```
    # Model
    model = Sequential()
    model.add(Dense(10, activation='relu',
        input_shape=(window_size,)))
    model.add(Dense(10, activation='relu'))
    model.add(Dense(1))
    ```

    Here, we are using three dense layers.

2. Next, we define our in-built schedule function:

    ```
    def inbuilt_schedule(model, train_data, X_val, y_val, optimizer,
    epochs=100):
        model.compile(optimizer=optimizer, loss='mse')
        model.fit(train_data, epochs=epochs, verbose=0)
        forecast_val = model.predict(X_val)
        mae_val, mse_val = evaluate_model(model, X_val, y_val)
        return mae_val, mse_val, forecast_val
    ```

    This function takes in the defined model, compiles it, trains the model with the training data (train_data) and evaluates it on the validation data (X_val, y_val). With this function we can pass in the optimizer along with the desired number of epochs we want to run the model for and it returns the evaluation metrics (mae and mse) along with the validation forecast as its output.

3. Let's set up our in-built learning rate schedules:

    ```
    # Set up learning rate schedules
    lr_schedules = {
        'Exponential Decay': tf.keras.optimizers.schedules.
    ```

```
ExponentialDecay(initial_learning_rate=0.1, decay_steps=100,
decay_rate=0.96),
    'Piecewise Constant': tf.keras.optimizers.schedules.
PiecewiseConstantDecay([30, 60], [0.1, 0.01, 0.001]),
    'Polynomial Decay': tf.keras.optimizers.schedules.
PolynomialDecay(initial_learning_rate=0.1, decay_steps=100, end_
learning_rate=0.01, power=1.0)
}
```

Lets discuss each of these schedules here:

I.   The exponential decay learning rate schedule sets up a learning rate that decays exponentially over time. In this experiment, the initial learning rate is set to 0.1. This learning rate will undergo an exponential decay at a rate of 0.96 for every 100 steps, as defined by the decay_steps parameter.

II.   The PiecewiseConstantDecay learning rate schedule allows us the flexibility to define specific learning rates for different periods during training. In our case, we specified 30 and 60 steps as our boundaries; this means that for the first 30 steps, we apply a learning rate of 0.1, from 30 to 60, we apply a learning rate of 0.01, and from 61 to the end of training, we apply a learning rate of 0.001. To PiecewiseConstantDecay, the number of learning rates should be one more than the number of boundaries applied. For example, in our case, we have two boundaries ([30, 60]) and three learning rates ([0.1, 0.01, 0.001]).

III. In this experiment, we set initial_learning_rate to 0.1; this serves as our starting learning rate. We set the decay_steps parameter to 100, indicating that the learning rate will decay over these 100 steps. Next, we set our end_learning_rate to 0.01; this means that by the conclusion of our decay_steps, the learning rate will have reduced to this value. The power parameter controls the exponent to which the step decay is raised. In this experiment, we set the power value to 1.0, resulting in linear decay.

4.   Next, let's try out the schedules:

```
forecasts = {}
results = {}

for i, lr_schedule in lr_schedules.items():
    optimizer = tf.keras.optimizers.Adam(learning_rate=lr_
schedule)
    mae, mse, forecast_val = inbuilt_scheduler(model, train_
data, X_val, y_val, optimizer)
    results[i] = (mae, mse)
```

```
        forecasts[i] = forecast_val

# Plot entire validation dataset first
for i, forecast_val in forecasts.items():
    # Plot predictions
    plot_predictions(forecast_val, y_val, title=i + " LR")

    # Print MAE and MSE values
    mae, mse = results[i]
    print(f"For {i} LR:")
    print(f"MAE: {mae}")
    print(f"MSE: {mse}")
    print("==========================")
```

Here we train our defined model using the three different schedules and evaluate each of them with the validation set, after which we plot the forecasts as shown in *Figure 13.1*:

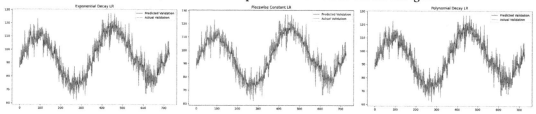

Figure 13.1 – Plots showing true forecast versus the validation
forecast using the three different learning schedules

From the plots its hard to distinguish which schedule closely follows the ground truth. Also the code block returns the MAE and MSE scores for each of the schedules used as shown in *Figure 13.2*.

	Learning Rate Schedule	**MAE**	**MSE**
1	Exponential Decay	4.24	28.29
2	Piecewise Constant	4.51	31.66
3	Polynomial Decay	4.21	27.991

Figure 13.2 – A table showing the learning rate schedule with the evaluation metrics.

We can see our `polynomialDecay` learning rate schedule won this battle, not only with the other learning rates schedules but also, it outperforms the simple DNN model with the same architecture that we used in *Chapter 12, Introduction to Time Series, Sequences, and Predictions*. You can read more about the learning rate schedule from the documentation `https://www.tensorflow.org/api_docs/python/tf/keras/callbacks/LearningRateScheduler` or using this excellent Medium article `https://rmoklesur.medium.com/learning-rate-scheduler-in-keras-cc83d2f022a6` by Moklesur Rahman.

1. We can zoom into specific dates to have a clearer view of how each schedule performs:

```
# Zoom into a specific date range
start_date = "2022-01-01"
end_date = "2022-06-30"

for i, forecast_val in forecasts.items():
    # Plot predictions for specific date range
    plot_predictions(forecast_val, y_val, title=i + " LR -
Zoomed In", start_date=start_date, end_date=end_date)
```

From the zoomed returned plots in *Figure 13.3* we see that the forecast using polynomial decay schedule is more closely aligned with the validation set in comparison to others.

Figure 13.3 – A zoomed in plot showing True forecast versus the validation
forecast using the three different learning schedules

Now you have a good idea of how to apply these learning rate schedule, you may want to tweak the values and see if you can achieve a much lower MAE and MSE. When you are done, let's look at a custom learning rate schedule.

Custom learning rate scheduler

Beyond using built-in learning rate schedule, TensorFlow provides us with an easy way to build a custom learning rate scheduler to help us find the optimal learning rate. Let's do this next:

1. Let's start by defining the custom learning rate scheduler:

```
# Define learning rate scheduler
lr_schedule = tf.keras.callbacks.LearningRateScheduler(
    lambda epoch: 1e-7 * 10**(epoch / 10))
```

Here, we start with a small learning rate (1×10^{-7}) and we increase this learning rate exponentially with each passing epoch. We use `10**(epoch / 10)` to determine the rate at which the learning rate increases.

2. We define the optimizer with the initial learning rate:

```
# Define optimizer with initial learning rate
optimizer = tf.keras.optimizers.SGD(
    learning_rate=1e-7)
```

Here, we used an SGD with a starting learning rate of 1×10^{-7}.

3. Next, we compile the model with the defined optimizer and set our loss as MSE:

```
model.compile(optimizer=optimizer, loss='mse')
```

4. Now, we train the model:

```
history = model.fit(train_data, epochs=200,
    callbacks=[lr_schedule], verbose=0)
```

We train the model for 200 epochs and then pass the learning rate scheduler as a callback. This way, the learning rate is adjusted based on the customization when defining our custom learning rate scheduler. We also set `verbose=0` so we don't print the training process.

5. Calculate the learning rates for each epoch:

```
lrs = 1e-7 * (10 ** (np.arange(200) / 10))
```

We use this code to calculate the learning rate per epoch and it gives us an array of learning rates.

6. We plot the model loss against the learning rate:

```
plt.semilogx(lrs, history.history["loss"])
plt.axis([1e-7, 1e-3, 0, 300])
plt.xlabel('Learning Rate')
plt.ylabel('Loss')
plt.title('Learning Rate vs Loss')
plt.show()
```

This plot is an effective way of selecting the optimal learning rate.

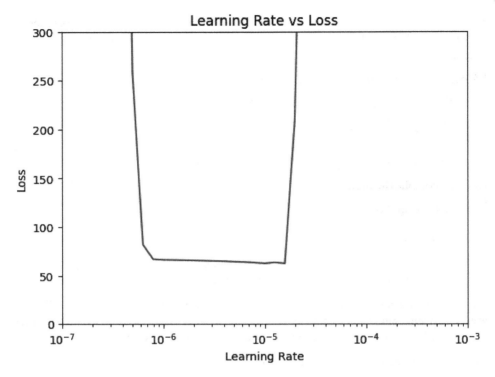

Figure 13.4 – The learning rate loss curve

To find the optimal learning rate, we are on the lookout for where the loss is decreasing most rapidly before it begins to increase again. From the plot, we can see the learning rate falls and settles at around 9×10^{-4}, after which it begins to rise again. So, we will pick this value as our ideal learning rate for this experiment. Now we will retrain our model using this new learning rate as our fixed learning rate. When we run the code, we get an MAE of 4.39 and an MSE of 30.11.

We have now seen how to use both in-built learning rate schedule and custom scheduler. Let us now switch our attention to using CNNs for time series forecasting.

CNNs for time series forecasting

CNNs have recorded remarkable success in image classification tasks due to their ability to detect localized patterns within grid-like data structures. This idea can also be applied to time series forecasting. By viewing a time series as a sequence of temporal intervals, CNNs can extract and recognize patterns that are predictive of future trends. Another important strength of CNNs is their translation-invariant nature. This means once they learn a pattern in one segment, the network is well equipped to recognize it everywhere else it occurs within the series. This comes in handy in detecting reoccurring patterns across time steps.

The setup of a CNN also helps to automatically reduce the dimensionality of our input data with the aid of the pooling layers. Hence, the convolution and pooling operations in a CNN transform the input series into a streamlined form that captures the core features while ensuring computational efficiency. Unlike with images, here we use a 1D convolutional filter because of the nature of time series data (singular dimension). This filter slides across the time dimension, observing localized windows of values as input. It detects informative patterns within these intervals through repeated element-wise multiplications and summations between its weights and the input windows.

Multiple filters are learned to extract diverse predictive signals – trends, seasonal fluctuations, cycles, and more. Similar to patterns within images, CNNs can recognize translated versions of these temporal motifs throughout the series. When we apply successive convolutional and pooling layers, the network composes these low-level features into higher-level representations, progressively condensing the series into its most salient components. Fully connected layers ultimately use these learned features to make forecasts.

Let us return to our notebooks and apply a 1D CNN in modeling our sales data. We already have our training and test data. Now, to model our data with a CNN, we need to carry out an extra step, which involves reshaping our data to meet the expected input shape a CNN expects. In *Chapter 7*, Image Classification with Convolutional Neural Networks, we saw how CNNs required 3D data in comparison to the 2D data we use when modeling with a DNN; the same applies here.

A CNN requires a batch size, a window size, and the number of features. The batch size is the first dimension of our input shape; it refers to the number of sequences we feed into the CNN. We set the window size value to 20, and the number of features here refers to the number of distinct features at each time step. For a univariate time series, this value will be 1; for a multivariate time series, the value will be 2 or more.

Since we are dealing with a univariate time series in our case study, our input shape needs to look like (128, 20, 1):

1. Let's prepare our data for modeling with a CNN with the right shape:

    ```
    # Make sequences
    window_size = 20
    X = []
    y = []

    for i in range(window_size, len(data)):
        X.append(data[i-window_size:i])
        y.append(data[i])

    X = np.array(X)
    y = np.array(y)

    # Train/val split
    split = int(0.8 * len(X))
    ```

```
X_train, X_val = X[:split], X[split:]
y_train, y_val = y[:split], y[split:]

# Reshape data
X_train = X_train.reshape(-1, window_size, 1)
X_val = X_val.reshape(-1, window_size, 1)

# Set batch size and shuffle buffer
batch_size = 128
shuffle_buffer = 1000

train_data = tf.data.Dataset.from_tensor_slices(
    (X_train, y_train))
train_data = train_data.shuffle(
    shuffle_buffer).batch(batch_size)
```

Most of the code in this code block is the same. The key step here is the reshape step, which we use to achieve the input shape required for CNN modeling.

2.   Let's build our model:

```
# Build model
model = Sequential()
model.add(Conv1D(filters=64, kernel_size=3,
    strides=1,
    padding='causal',
    activation='relu',
    input_shape=(window_size, 1)))

model.add(MaxPooling1D(pool_size=2))

model.add(Conv1D(filters=32, kernel_size=3,
    strides=1,
    padding='causal',
    activation='relu'))

model.add(MaxPooling1D(pool_size=2))
model.add(Flatten())
model.add(Dense(16, activation='relu'))
model.add(Dense(1))
```

In our model, we apply 1D convolutional layers due to the single-dimensional nature of time series data, unlike the 2D CNNs we employed for image classification, due to the 2D structure of images. Here, our model is made up of two 1D convolutional layers, each followed by a max pooling layer. In our first convolutional layer, we have 64 filters to learn various data patterns with a filter size of 3, which allows it to recognize patterns spanning three-time steps. We use a stride of 1; this means that our filter traverses the data one step at a time, and to ensure nonlinearity, we use ReLU as our activation function. Notice that we are using a new type of padding, called `causal padding`. This choice is strategic as causal padding ensures that the model's output for a particular time step is influenced only by that time step and its predecessors, never by future data. By adding padding to the start of the sequence, causal padding respects the natural temporal sequence of our data. This is important to prevent our model from inadvertently "looking ahead," ensuring forecasts rely solely on past and current information.

We earlier outlined that we need 3D input-shaped data to be fed into our CNN model made up of the batch size, window size, and number of features. Here, we used `input_shape=(window_size, 1)`. We did not state the batch size in the input shape definition. This means the model can take batches of different sizes since we did not hardcode any batch size. Also, we only have one feature since we are dealing with a univariate time series, and that's why we have specified 1 in the input shape along with the window size. The max pooling layer reduces the dimensionality of our data. Next, we reach the second convolutional layer; this time we use 32 filters, again with a kernel size of 3, causal padding, and ReLU as the activation function. Next, the max pooling layer samples the data again. After this, the data is flattened and fed into the fully connected layers to make predictions based on the patterns learned from our sales data.

3. Let's compile and fit the model for 100 epochs:

```
model.compile(loss='mse', optimizer='adam')
# Train model
model.fit(train_data, epochs=100)
```

4. Finally, let's evaluate the performance of our model:

```
# evaluate
mae, mse = evaluate_model(model, X_val_cnn, y_val)
print(f'MAE: {mae}, MSE: {mse}')
```

We evaluate the model on the validation set by generating the MAE and MSE. When we run the code, we achieve an MAE of 5.50 and MSE of 47.44. Here, you have the opportunity to try achieve a much lower MAE by tweaking the model's hyperparameters, this can serve as a good exercise for you.

Next, let us see how we can use the RNN family in forecasting time series data.

RNNs in time series forecasting

Time series forecasting poses a unique challenge in the world of machine learning, involving the prediction of future values based on previously observed sequential data. An intuitive way of thinking about this is to consider a sequence of past data points. The question then becomes, given this sequence, how can we predict the next data point or sequence of data points? This is where RNNs demonstrate their efficacy. RNNs are a specific type of neural network developed to process sequential data. They maintain an internal state or "memory" that holds information about the elements of the sequence observed thus far. This internal state is updated at each step of the sequence, merging information from the new input and the previous state. As an example, while predicting sales, an RNN may retain data regarding the sales trends from the previous months, the overall trend across the past year, and the seasonal effects, among others.

However, standard RNNs exhibit a significant limitation: the problem of "vanishing gradients." This problem results in difficulty in maintaining and utilizing information from earlier steps in the sequence, especially as the sequence length increases. To overcome this hurdle, the deep learning community introduced advanced architectures. LSTMs and GRUs are specialized types of RNNs designed explicitly to counteract the vanishing gradient issue. These types of RNNs are capable of learning long-term dependencies due to their in-built gating mechanisms, which control the flow of information in and out of the memory state.

Thus, RNNs, LSTMs, and GRUs can be potent tools for time series forecasting because they inherently incorporate the temporal dynamics of the problem. For instance, while predicting sales, these models can factor in seasonal patterns, holidays, weekends, and more by maintaining information about previous sales periods, which could lead to more accurate forecasts.

Let's put a simple RNN into action here and see how it will perform on our dataset:

1. Let's start by preparing our data:

```
# Make dataset
batch_size = 128
dataset = tf.data.Dataset.from_tensor_slices((X_train, y_train))
#we use X_train not X_train_cnn
dataset = dataset.shuffle(buffer_size=1024).batch(batch_size)
```

Here, you will observe we are preparing our data in the same way as we did when using the DNN and we do not reshape it as we just did with our CNN model. There is a simple trick ahead that will help you do this in our model architecture.

2. Let's define our model architecture:

```
model = tf.keras.models.Sequential([
    tf.keras.layers.Lambda(lambda x: tf.expand_dims(
        x, axis=-1),
        input_shape=[None]),
    tf.keras.layers.SimpleRNN(40,
        return_sequences=True),
    tf.keras.layers.SimpleRNN(40),
    tf.keras.layers.Dense(1),
    tf.keras.layers.Lambda(lambda x: x * 100.0)
])
```

When modeling with RNNs, just like we saw with CNNs, we need to reshape our data as our model expects 3D-shaped input data as well. However, in scenarios where you would like to keep your original input shape intact for various experiments with different models, we can resort to a simple yet effective solution – the Lambda layer. This layer is a powerful tool in our toolbox that lets us perform simple, arbitrary functions on the input data, making it an excellent instrument for quick experimentation.

With Lambda layers, we can execute element-wise mathematical operations, such as normalization, linear scaling, and simple arithmetic operations. In our case, we utilize a Lambda layer to expand the dimension of our 2D input data to fit the 3D input requirement (batch_size, time_steps, and features) of RNNs. In TensorFlow, you can leverage the Keras API's tf.keras.layers.Lambda to create a Lambda layer. A Lambda layer serves as an adapter, allowing us to make tweaks to the data, ensuring it's in the right format for our model, while keeping the original data intact for other uses. Next, we come across two simple RNN layers of 40 units each. It is important to note that in the first RNN, we included return_sequence =True. We use this in RNNs and LSTMs when the output of one RNN or LSTM layer is fed into another RNN or LSTM layer. We set this parameter to ensure the first RNN layer will return an output for each input in the sequence. The output is then fed into the second RNN layer; this layer will only return the output of the final step, which is then fed into the dense layers, which outputs the predicted value for each sequence. Then, we come across another Lambda layer that multiplies the output by 100. We use this to expand the output values.

> **Note**
> When we use architectures such as CNNs, RNNs, LSTMs, and CNN-LSTM, one approach is to reshape the data during the pre-processing step before we feed it into the model, or we reshape our input data with the aid of lambda layers within the model.

3. Let's compile and fit our model:

```
# Compile model
model.compile(optimizer=tf.keras.optimizers.Adam(learning_
rate=8e-4), loss='mse')
# Train model
model.fit(dataset, epochs=100)
```

Here, we achieve an MAE of 4.29 and an MSE of 28.70 on the validation set. This result is slightly worse than our best result so far. Perhaps here, you have an opportunity to try our different learning rate schedules, may be a lower MAE is on the horizon, give it a try. Next, let us explore LSTMs.

LSTMs in time series forecasting

In the NLP section, we discussed the capabilities of LSTMs and their improvements over RNNs by mitigating issues such as the vanishing gradient problem, enabling the model to learn longer sequences. In the context of time series forecasting, LSTM networks can be quite powerful. Let's see how we can apply LSTMs to our sales dataset:

1. Let's define our model: In this experiment we reshape our data like we did when we built a CNN architecture, hence we do not use the Lambda layers to reshape the input.

    ```
    model_lstm = tf.keras.models.Sequential([
        tf.keras.layers.LSTM(50, return_sequences=True,
            input_shape=[None, 1]),
        tf.keras.layers.LSTM(50),
        tf.keras.layers.Dense(1)
    ])
    ```

    The first layer is an LSTM layer of 50 neurons, and it has the `return_sequence` parameter set to `True` to ensure the output returned is the complete sequence, which is fed into the final LSTM layer. Here, we also use an input shape of `[None, 1]`. The next layer also has 50 neurons, this outputs a single value at each time step since we did not set the `return_sequence` parameter to `True` here and this is fed into the dense layer for predictions.

2. The next step is to compile the model. We compile and fit our model as before, then we evaluate it. Here, we achieved an MAE of 4.52 and an MSE of 32.45, this.

Let's see if we can improve on our best result using a CNN-LSTM architecture, let try this next.

CNN-LSTM architecture for time series forecasting

Deep learning has offered compelling solutions for time series forecasting, and one of the notable architectures in this space is the CNN-LSTM model. This model leverages the strengths of CNNs and LSTM networks, providing an effective framework for handling the unique characteristics of time series data. CNNs are renowned for their performance in image processing tasks due to their ability to learn spatial patterns in images, while in sequential data, they can learn local patterns. The convolutional layers within the network apply a series of filters to the data, learning and extracting significant local and global temporal patterns and trends. These features act as a compressed representation of the original data, retaining essential information while reducing dimensionality. The reduction in dimensionality could lead to a more efficient representation that captures relevant patterns.

Once significant features have been extracted through the convolutional layers, these features become inputs to the LSTM layer(s) of the network. CNN-LSTM models have an advantage in their capacity for end-to-end learning. In this architecture, CNN and LSTM have complementary roles. The CNN layer captures local patterns, and the LSTM layers learn temporal relationships from these patterns.

This joint optimization is central to the performance gains of the CNN-LSTM model in comparison to independent architectures. Let us see how to apply this architecture to our sales data. Here we are going straight to the model architecture as we have already looked at the data preparation steps several times before:

1. Let's build our model using convolution layers:

    ```
    # Build the Model
    model = tf.keras.models.Sequential([
        tf.keras.layers.Conv1D(filters=64, kernel_size=3,
        strides=1,
        activation="relu",
        padding='causal',
        input_shape=[window_size, 1]),
    ```

    We use our 1D convolutional layer to detect patterns from the sequence of values as we did in our CNN forecasting experiment. Remember to set `padding` to `causal` to ensure the output size remains the same as the input. We set the input shape to `[window_size, 1]`. Here, `window_size` represents the number of time steps in each input sample. `1` means we are working with a univariate time series. For example, if we set `window_size` to 7, this will mean we are using a week's worth of data for forecasting.

2. Next, our data reaches the LSTM layers, which are made up of 2 LSTM layers, each made up of 64 neurons:

    ```
    tf.keras.layers.LSTM(64, return_sequences=True),
    tf.keras.layers.LSTM(64),
    ```

3. Then, we have the dense layers:

    ```
    tf.keras.layers.Dense(30, activation="relu"),
    tf.keras.layers.Dense(10, activation="relu"),
    tf.keras.layers.Dense(1),
    ```

    The LSTM layer adds temporal context to the features extracted by the CNN layers, and the dense layers generate the final predictions. Here, we use three fully connected layers and output the final forecasted values. With this architecture, we achieve an MAE of 4.78 and an MSE of 35.17.

Tuning the hyperparameters to optimize our model's performance is a good idea here. By adjusting the hyperparameters such as the learning rate, batch size, or optimizers, we may be able to improve the model's capabilities. We will not be going into this here as you are already well equipped to do this. Let us move on to the Apple stock price data and use all we have learned to create a series of experiments to forecast the future prices of Apple stocks and see which architecture will come out on top.

Forecasting Apple stock price data

We have now covered everything we need to know about time series for the TensorFlow Developer Certificate exam. Let us round off this chapter and the book with a real-world use case on time series. For this exercise, we will be working with a real-world data set (Apple closing day stock price). The Jupyter notebook for this exercise can be found here: https://github.com/PacktPublishing/TensorFlow-Developer-Certificate-Guide/blob/main/Chapter%2013/Apple_Stock_forecasting.ipynb.

1. We start by importing the required libraries:

```
import pandas as pd
import numpy as np
import matplotlib.pyplot as plt
import tensorflow as tf
from tensorflow import keras
import yfinance as yf
from sklearn.metrics import mean_absolute_error, mean_squared_
error
from tensorflow.keras.models import Sequential
from tensorflow.keras.layers import Dense, Conv1D, MaxPooling1D,
Flatten, Lambda, SimpleRNN, LSTM
```

Here, we are using a new library called yfinance. This lets us access the Apple stock data for our case study.

> **Note**
>
> You may want to run pip install yfinance to get it working if the import fails.

2. Create a DataFrame:

```
# Load the Apple closing stock price from Yahoo Finance
tickerSymbol = 'AAPL'
df = yf.Ticker(tickerSymbol)
df = df.history(period='1d', start='2013-01-01', end='2023-01-
01')
```

We set up a DataFrame using the AAPL as the ticker symbol, which represents Apple (the company) in the stock market. To do this, we use the yf.Ticker function from the yfinance library to access Apple's historical data from Yahoo Finance. We apply the history method to our Ticker object to access the historical market data for Apple. Here, we set the period to 1d, which means daily data should be accessed. We also set the start and end parameters to define the date range we want to access; in this case, we are collecting 10 years of data from the first day of January 2013 to the last day of January 2023.

3. Next, we use `df.head()` to get a snapshot of our DataFrame. We can see the dataset is made up of seven columns (`Open`, `High`, `Low`, `Close`, `Volume`, `Dividends`, and `Stock Splits`), as shown in *Figure 13.5*.

Date	Open	High	Low	Close	Volume	Dividends	Stock Splits
2013-01-02 00:00:00-05:00	16.960551	16.996688	16.587237	16.813858	560518000	0.0	0.0
2013-01-03 00:00:00-05:00	16.778644	16.833462	16.567946	16.601633	352965200	0.0	0.0
2013-01-04 00:00:00-05:00	16.444534	16.495371	16.103375	16.139206	594333600	0.0	0.0
2013-01-07 00:00:00-05:00	15.986078	16.209638	15.777831	16.044266	484156400	0.0	0.0
2013-01-08 00:00:00-05:00	16.206877	16.288951	15.963106	16.087440	458707200	0.0	0.0

Figure 13.5 – A snapshot of the Apple stock data

Let's understand what these columns mean:

- `Open` stands for the opening price for the trading day.

- `High` is the highest price at which stocks were traded during the day.

- `Low` is the lowest price at which stocks were traded during the day.

- `Close` stands for the closing price for the trading day.

- `Volume` signifies the number of shares that changed hands during the course of the trading day. This can serve as an indicator of the market strength.

- `Dividends` represents how the company's earnings are shared among shareholders.

- `Stock Splits` can be viewed as an act of the corporation that increased the number of the company's outstanding shares by splitting each share.

4. Let's plot the daily closing price:

```
plt.figure(figsize=(14,7))
plt.plot(df_apple.index, df_apple['Close'],
    label='Close price')
plt.title('Historical prices for AAPL')
plt.xlabel('Date')
plt.ylabel('Price')
plt.grid(True)
plt.legend()
plt.show()
```

When we run the code, we get the following plot:

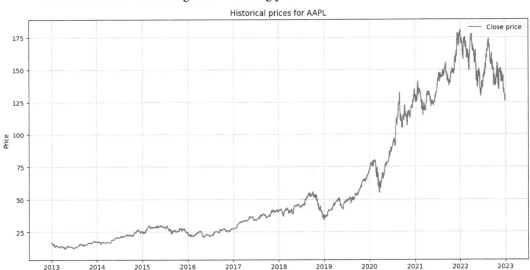

Figure 13.6 – A plot showing the Apple stock closing price between January 2013 and January 2023

From the plot in *Figure 13.6*, we can see that the stock has a positive upward trend with occasional dips.

5. Convert the data into a NumPy array:

```
data = df_apple['Close'].values
```

For this exercise, we will be forecasting the daily stock closing price. Hence, we take the closing price column and convert it into a NumPy array. This way, we create a univariate time series for our experimentation.

6. Prepare the windowed dataset:

```
# Sliding window,
window_size = 20
X, y = [], []
for i in range(window_size, len(data)):
    X.append(data[i-window_size:i])
    y.append(data[i])

X = np.array(X)
y = np.array(y)

# Train/val split
```

```
train_size = int(len(X) * 0.8)
X_train, X_val = X[:train_size], X[train_size:]
y_train, y_val = y[:train_size], y[train_size:]

# Dataset using tf.data
batch_size = 128
buffer_size = 1000
train_data = tf.data.Dataset.from_tensor_slices(
    (X_train, y_train))
train_data = train_data.cache().shuffle(
    buffer_size).batch(batch_size).prefetch(
    tf.data.experimental.AUTOTUNE)
```

We are now familiar with this code block, which we use to prepare our data for modeling. Next, we will use the same set of architectures to carry out our experiment with our Apple stock dataset. We can see the results in *Figure 13.7*:

Model	MAE
DNN	6.02
CNN	5.35
RNN	2.37
LSTM	2.61
CNN-LSTM	19.33

Figure 13.7 – A table showing the MAE across various models

From our results. we see that we achieved the best-performing model with our RNN architecture, having an MAE of 2.37. You can now save your best model, which can be used to predict future stock values. You can also tweak the hyperparameters further to see whether you can achieve a lower MAE.

> **Note**
>
> This model is an illustration of what is possible with forecasting with neural networks. However, we must be aware of its limitations. Please refrain from making financial decisions using this model as real-world stock market predictions capture complex relationships such as economic indicators, market sentiment, and other interdependencies, which our basic model does not take into consideration.

With this, we have come to end of this chapter and the book.

Summary

In this final chapter, we explored some advanced concepts for working with time series forecasting with TensorFlow. We saw both how to use in-built learning rate schedules as well as designing custom-made scheduler. Then, we used more specialized models, such as RNNs, LSTM networks, and the combination of CNNs and LSTM. We also saw how we could apply Lambda layers to implement custom operations and add flexibility to our network architecture.

To conclude the chapter, we worked on forecasting the Apple stock closing price. By the end of this chapter, you should have a good understanding of how to apply concepts such as learning rate schedules and Lambda layers and effectively building time series forecasting models using various architectures in readiness for your exam. Good luck!

Note from the author

It gives me great pleasure to see you move from the very fundamentals of machine learning to building various projects using TensorFlow. You have now explored building models with different neural network architectures. You now have a solid foundation upon which you can, and should, build an incredible career as a certified TensorFlow developer. You can only become a top developer by building solutions.

Everything we have covered in this book will serve you well as you finalize your preparation for the TensorFlow Developer Certificate exam and beyond. I want to congratulate you for not giving up and working through all the concepts and projects in this book and the exercises. I would like to encourage you to continue learning, experimenting, and keeping tabs on the latest developments in the field of machine learning. I wish you success in your exam and your career ahead.

Questions

1. Load the Google stock data from Yahoo Finance for 01-01-2015 to 01-01-2020.
2. Create training, forecasting, and plotting functions.
3. Prepare the data for training.
4. Build DNN, CNN, LSTM, and CNN-LSTM models to model the data.
5. Evaluate the models using MAE and MSE.

References

- Shi, X., Chen, Z., Wang, H., Yeung, D. Y., Wong, W. K., & Woo, W. C. (2015). *Convolutional LSTM Network: A Machine Learning Approach for Precipitation Nowcasting*. In Advances in Neural Information Processing Systems (pp. 802–810) https://papers.nips.cc/paper/2015/hash/07563a3fe3bbe7e3ba84431ad9d055af-Abstract.html

- Karim, F., Majumdar, S., Darabi, H., & Harford, S. (2019). *LSTM fully convolutional networks for time series classification.* IEEE Access, 7, 1662-1669

- Siami-Namini, S., Tavakoli, N., & Siami Namin, A. (2019). *The Performance of LSTM and BiLSTM in Forecasting Time Series.* In 2019 IEEE International Conference on Big Data

- TensorFlow learning rate scheduler: `https://www.tensorflow.org/api_docs/python/tf/keras/callbacks/LearningRateScheduler`

- *Lambda layers*: `https://keras.io/api/layers/core_layers/lambda/`

- *Time series forecasting*: `https://www.tensorflow.org/tutorials/structured_data/time_series`

- *tf.data: Build TensorFlow input pipelines.* `https://www.tensorflow.org/guide/data`

- *Windowed datasets for time series*: `https://www.tensorflow.org/tutorials/structured_data/time_series#data_windowing`

Index

Packtpub.com

Subscribe to our online digital library for full access to over 7,000 books and videos, as well as industry leading tools to help you plan your personal development and advance your career. For more information, please visit our website.

Why subscribe?

- Spend less time learning and more time coding with practical eBooks and Videos from over 4,000 industry professionals

- Improve your learning with Skill Plans built especially for you

- Get a free eBook or video every month

- Fully searchable for easy access to vital information

- Copy and paste, print, and bookmark content

Did you know that Packt offers eBook versions of every book published, with PDF and ePub files available? You can upgrade to the eBook version at packtpub.com and as a print book customer, you are entitled to a discount on the eBook copy. Get in touch with us at customercare@packtpub.com for more details.

At www.packtpub.com, you can also read a collection of free technical articles, sign up for a range of free newsletters, and receive exclusive discounts and offers on Packt books and eBooks.

Other Books You May Enjoy

If you enjoyed this book, you may be interested in these other books by Packt:

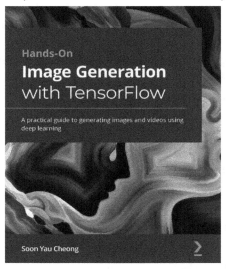

Hands-On Image Generation with TensorFlow

Soon Yau Cheong

ISBN: 9781838826789

- Train on face datasets and use them to explore latent spaces for editing new faces
- Get to grips with swapping faces with deepfakes
- Perform style transfer to convert a photo into a painting
- Build and train pix2pix, CycleGAN, and BicycleGAN for image-to-image translation
- Use iGAN to understand manifold interpolation and GauGAN to turn simple images into photorealistic images
- Become well versed in attention generative models such as SAGAN and BigGAN
- Generate high-resolution photos with Progressive GAN and StyleGAN

TensorFlow 2.0 Computer Vision Cookbook

Jesús Martínez

ISBN: 9781838829131

- Understand how to detect objects using state-of-the-art models such as YOLOv3
- Use AutoML to predict gender and age from images
- Segment images using different approaches such as FCNs and generative models
- Learn how to improve your network's performance using rank-N accuracy, label smoothing, and test time augmentation
- Enable machines to recognize people's emotions in videos and real-time streams
- Access and reuse advanced TensorFlow Hub models to perform image classification and object detection
- Generate captions for images using CNNs and RNNs

Packt is searching for authors like you

If you're interested in becoming an author for Packt, please visit authors.packtpub.com and apply today. We have worked with thousands of developers and tech professionals, just like you, to help them share their insight with the global tech community. You can make a general application, apply for a specific hot topic that we are recruiting an author for, or submit your own idea.

Share Your Thoughts

Now you've finished *TensorFlow Developer Certificate Guide*, we'd love to hear your thoughts! Scan the QR code below to go straight to the Amazon review page for this book and share your feedback or leave a review on the site that you purchased it from.

https://packt.link/r/1-803-24013-X

Your review is important to us and the tech community and will help us make sure we're delivering excellent quality content.

Download a free PDF copy of this book

Thanks for purchasing this book!

Do you like to read on the go but are unable to carry your print books everywhere?

Is your eBook purchase not compatible with the device of your choice?

Don't worry, now with every Packt book you get a DRM-free PDF version of that book at no cost.

Read anywhere, any place, on any device. Search, copy, and paste code from your favorite technical books directly into your application.

The perks don't stop there, you can get exclusive access to discounts, newsletters, and great free content in your inbox daily

Follow these simple steps to get the benefits:

1. Scan the QR code or visit the link below

https://packt.link/free-ebook/978-1-80324-013-8

2. Submit your proof of purchase
3. That's it! We'll send your free PDF and other benefits to your email directly

www.ingramcontent.com/pod-product-compliance
Lightning Source LLC
Chambersburg PA
CBHW080620060326
40690CB00021B/4760